# Artificial Intelligence in Society

This document, as well as any data and any map included herein, are without prejudice to the status of or sovereignty over any territory, to the delimitation of international frontiers and boundaries and to the name of any territory, city or area.

**Please cite this publication as:**

OECD (2019), *Artificial Intelligence in Society*, OECD Publishing, Paris, *https://doi.org/10.1787/eedfee77-en*.

ISBN 978-92-64-58254-5 (print)
ISBN 978-92-64-54519-9 (pdf)

The statistical data for Israel are supplied by and under the responsibility of the relevant Israeli authorities. The use of such data by the OECD is without prejudice to the status of the Golan Heights, East Jerusalem and Israeli settlements in the West Bank under the terms of international law.

**Photo credits:** Cover © Adobe Stock.

Corrigenda to OECD publications may be found on line at: *www.oecd.org/about/publishing/corrigenda.htm*.

# *Preface*

Artificial intelligence (AI) is reshaping economies, promising to generate productivity gains, improve efficiency and lower costs. It contributes to better lives and helps people make better predictions and more informed decisions. These technologies, however, are still in their infancy, and there remains much promise for AI to address global challenges and promote innovation and growth. As AI's impacts permeate our societies, its transformational power must be put at the service of people and the planet.

At the same time, AI is also fuelling anxieties and ethical concerns. There are questions about the trustworthiness of AI systems, including the dangers of codifying and reinforcing existing biases, such as those related to gender and race, or of infringing on human rights and values, such as privacy. Concerns are growing about AI systems exacerbating inequality, climate change, market concentration and the digital divide. No single country or actor has all the answers to these challenges. We therefore need international co-operation and multi-stakeholder responses to guide the development and use of AI for the wider good.

This book, *Artificial Intelligence in Society*, examines the AI landscape and highlights key policy questions. Its goal is to help build a shared understanding of AI in the present and near term, and to encourage a broad dialogue on important policy issues, such as labour market developments and upskilling for the digital age; privacy; accountability of AI-powered decisions; and the responsibility, security and safety questions that AI generates.

The book draws on the work of the AI group of experts at the OECD, formed in 2018 to scope principles to facilitate innovation, adoption and trust in AI. Their debates inspired the OECD *Recommendation of the Council on Artificial Intelligence* – the first intergovernmental standard on AI – adopted by all OECD members and by several partner countries on 22 May 2019. This work emphasises the need for international co-operation to shape a policy environment that fosters trust in and adoption of AI.

Looking ahead, we must progress together on AI-related technical, ethical and legal issues, in order to foster the alignment of standards and codes of conduct while ensuring the inter-operability of laws and regulations. This is urgent, given the speed of developments and the breadth of applications. It is thus no surprise that AI is a top priority on national and international agendas including in the G7 and G20.

Adoption of the Recommendation and creation of a global dialogue are vital first steps. But there is much more to be done. With the establishment of the OECD AI Policy Observatory later this year, we are bringing our analytical, measurement and policy expertise to bear on largely uncharted territory. The Observatory – an inclusive hub for public policy on AI – will help countries encourage, nurture and monitor the responsible development of trustworthy AI systems for the benefit of society.

Moving forward, the OECD is gearing up to move from principles to action. We are determined to help countries implement the Recommendation to ensure that our societies and economies harness the full promise of AI, sharing its benefits broadly and putting the right safeguards in place so that no one is left behind –now and for generations to come.

Angel Gurría

Secretary-General

OECD

# *Foreword*

This book aims to help build a shared understanding of artificial intelligence (AI) in the present and near term. The book maps the economic and social impacts of AI technologies and applications and their policy implications, presenting evidence and policy options. It is also intended to help co-ordination and consistency with discussions in other international fora, notably the G7, the G20, the European Union and the United Nations.

The book builds on the OECD October 2017 Conference "AI: Intelligent Machines, Smart Policies" (http://oe.cd/ai2017); on the activities and discussions of the AI Group of experts at the OECD (AIGO) from September 2018 through February 2019; and on the OECD *Recommendation of the Council on Artificial Intelligence*. In turn, it has contributed to the OECD Going Digital project and the OECD's publication *Going Digital: Shaping Policies, Improving Lives*.

Chapter 1, "The AI technical landscape", provides a historical overview of AI's evolution from the development of symbolic AI in the 1950's to recent achievements in machine learning. It presents the work of the OECD's AI Group of Experts (AIGO) to describe AI systems – that predict, recommend or decide an outcome to influence the environment – and their lifecycle. The chapter also proposes a research taxonomy to help policy makers understand AI trends and identify policy issues.

Chapter 2, "The AI economic landscape", discusses AI's role as a new general-purpose technology that can lower the cost of prediction and enable better decisions. Complementary investments in data, skills and digitalised workflows are required, as is the capacity to adapt organisational processes. The chapter also reviews trends in private equity investment in AI start-ups.

Chapter 3, "AI applications", considers ten areas that are experiencing rapid uptake of AI technologies – transport, agriculture, finance, marketing and advertising, science, health, criminal justice, security, the public sector and augmented/virtual reality. Benefits of AI use in these areas include improving the efficiency of decision making, saving costs and enabling better resource allocation.

Chapter 4, "Public policy considerations", reviews salient policy issues that accompany the diffusion of AI. The chapter supports the OECD AI Principles adopted in May 2019, first in terms of values: inclusive growth, sustainable development and well-being; human-centred values and fairness; transparency and explainability; robustness, security and safety; and accountability. Secondly, it outlines national policies to promote trustworthy AI systems: investing in responsible AI research and development; fostering a digital ecosystem for AI; shaping an enabling policy environment for AI; preparing people for job transformation and building skills; and measuring progress.

Chapter 5, "AI policies and initiatives", illustrates the growing importance of AI in the policy agendas of stakeholders at both national and international levels. All stakeholder groups – governments and inter-governmental organisations, as well as companies, technical organisations, academia, civil society and trade unions – are actively engaged in discussions on how to steer AI development and deployment to serve all of society.

This book was declassified by the OECD Committee on Digital Economy Policy (CDEP) on 10 April 2019 by written procedure and prepared for publication by the OECD Secretariat.

# *Acknowledgements*

The *Artificial Intelligence in Society* publication was prepared under the aegis of the OECD Committee for Digital Economy Policy (CDEP), with input from its working parties. CDEP delegates contributed significantly with their comments and amendments, as well as by sharing and reviewing their countries' national AI strategy.

The core authors of the publication were Karine Perset, Nobuhisa Nishigata and Luis Aranda of the OECD's Digital Economy Policy Division, with Karine Perset conducting overall editing and co-ordination. Anne Carblanc, Head of the OECD Digital Economy Policy Division; Andrew Wyckoff and Dirk Pilat, respectively Director and Deputy Director for Science, Technology and Innovation, provided leadership and oversight. Parts of the book were researched and drafted by Doaa Abu Elyounes, Gallia Daor, Lawrence Pacewicz, Alistair Nolan, Elettra Ronchi, Carlo Menon, Christian Reimsbach-Kounatze. Guidance, input and comments were provided by experts across the OECD, including Laurent Bernat, Dries Cuijpers, Marie-Agnes Jouanjean, Luke Slawomirski, Mariagrazia Squicciarini, Barbara Ubaldi and Joao Vasconcelos.

The book benefited from the contributions of Taylor Reynolds and Jonathan Frankle from the MIT Internet Policy Research Initiative; Douglas Frantz, independent consultant; Avi Goldfarb from the University of Toronto; Karen Scott from Princeton University; the OECD's Trade Union Advisory Committee; Amar Ashar, Ryan Budish, Sandra Cortesi, Finale Doshi-Velez, Mason Kortz and Jessi Whitby, from the Berkman Klein Center for Internet and Society at Harvard University; and the members of the AI Group of Experts at the OECD (AIGO). The drafting team wishes in particular to thank Nozha Boujemaa, Marko Grobelnik, James Kurose, Michel Morvan, Carolyn Nguyen, Javier Juárez Mojica and Matt Chensenfor their valuable inputs and feedback.

The book also leverages ongoing work streams throughout the OECD. These include work by the Committee for Scientific and Technological Policy and its Working Party on Innovation and Technology Policy; the Committee on Consumer Policy and its Working Party on Consumer Product Safety; the Committee on Industry, Innovation and Entrepreneurship and its Working Party on Industrial Analysis; the Employment, Labour and Social Affairs Committee; the Education Policy Committee; and the e-leaders initiative of the Public Governance Committee and the Competition Committee, in addition to the Committee on Digital Economy Policy and its working parties, notably the Working Party on Security and Privacy in the Digital Economy.

The authors are also grateful to Mark Foss for editing this publication, and to Alice Weber and Angela Gosmann for editorial support. The publication's overall quality benefited from their engagement.

The support of the Japanese Ministry for Information and Communications (MIC) for this project is gratefully acknowledged.

# Table of contents

## Tables

## Figures

## Boxes

*Acronyms, abbreviations and currencies*

| | |
|---|---|
| AGI | Artificial general intelligence |
| AI | Artificial intelligence |
| AI HLEG | High-Level Expert Group on AI (European Commission) |
| AIGO | AI Group of Experts (OECD) |
| AINED | Dutch public-private partnership "AI for the Netherlands" |
| AIS | Autonomous and intelligent system |
| ANI | Artificial narrow intelligence |
| AR | Augmented reality |
| AUD | Australian dollar |
| AV | Autonomous vehicle |
| CAD | Canadian dollar |
| CHF | Swiss franc |
| CoE | Council of Europe |
| CTR | Click-through rate |
| DKK | Danish krone |
| EC | European Commission |
| EESC | European Economic and Social Committee |
| EHR | Electronic health record |
| EUR | Euro |
| FICO | Fair, Isaac and Company |
| G20 | Group of Twenty |
| G7 | Group of Seven |
| GBP | British pound |
| GDPR | General Data Protection Regulation (European Union) |
| GM | General Motors |
| HoME | Household Multimodal Environment |
| HRIA | Human rights impact assessment |
| IEC | International Electrotechnical Commission |

| | |
|---|---|
| IEEE | Institute for Electrical and Electronics Engineers |
| IoT | Internet of Things |
| IP | Intellectual property |
| IPRs | Intellectual property rights |
| IPRI | Internet Policy Research Initiative (Massachusetts Institute of Technology) |
| ISO | International Organization for Standardization |
| ITI | Information Technology Industry Council |
| KRW | Korean won |
| MIT | Massachusetts Institute of Technology |
| ML | Machine learning |
| MPC | Secure multi-party computation |
| MPI | Max Planck Institute for Innovation and Competition |
| NLP | Natural language processing |
| NOK | Norwegian krone |
| OAI | Office for Artificial Intelligence (United Kingdom) |
| PAI | Partnership on Artificial Intelligence to Benefit People and Society |
| PIAAC | Programme for the International Assessment of Adult Competencies (OECD) |
| R&D | Research and development |
| RMB | Yuan renminbi |
| SAE | Society of Automotive Engineers |
| SAR | Saudi riyal |
| SDGs | Sustainable Development Goals (United Nations) |
| SMEs | Small and medium-sized enterprises |
| STEM | Science, technology, engineering and mathematics |
| UGAI | Universal Guidelines on Artificial Intelligence |
| USD | US dollar |
| VR | Virtual reality |

## *Executive summary*

### Machine learning, big data and computing power have enabled recent AI progress

The artificial intelligence (AI) technical landscape has evolved significantly from 1950 when Alan Turing first posed the question of whether machines can think. Coined as a term in 1956, AI has evolved from symbolic AI where humans built logic-based systems, through the AI "winter" of the 1970s to the chess-playing computer Deep Blue in the 1990s. Since 2011, breakthroughs in "machine learning" (ML), an AI subset that uses a statistical approach, have been improving machines ability to make predictions from historical data. The maturity of a ML modelling technique called "neural networks", along with large datasets and computing power, is behind the expansion in AI development.

### AI systems predict, recommend or decide an outcome to influence the environment

An AI system, as explained by the OECD's AI Experts Group (AIGO), is a

*machine-based system that can, for a given set of human-defined objectives, make predictions, recommendations or decisions influencing real or virtual environments. It uses machine and/or human-based inputs to perceive real and/or virtual environments; abstract such perceptions into models (in an automated manner e.g. with ML or manually); and use model inference to formulate options for information or action. AI systems are designed to operate with varying levels of autonomy.*

The AI system lifecycle phases are i) planning and design, data collection and processing, and model building and interpretation; ii) verification and validation; iii) deployment; and iv) operation and monitoring. An AI research taxonomy distinguishes AI applications, e.g. natural language processing; techniques to teach AI systems, e.g. neural networks; optimisation, e.g. one-shot-learning; and research addressing societal considerations, e.g. transparency.

### AI can improve productivity and help solve complex problems

The AI economic landscape is evolving as AI becomes a general-purpose technology. Through cheaper and more accurate predictions, recommendations or decisions, AI promises to generate productivity gains, improve well-being and help address complex challenges. Leveraging AI requires complementary investments in data, skills and digitalised workflows, as well as changes to organisational processes. Therefore, adoption varies across companies and industries.

### AI investment and business development are growing rapidly

Private equity investment in AI start-ups accelerated from 2016, after five years of steady increases. Private equity investment doubled from 2016 to 2017, reaching USD 16 billion in 2017. AI start-ups attracted 12% of worldwide private equity investments in the first half of 2018, reflecting a significant increase from just 3% in 2011 in a trend seen across all major economies. These investments are usually large, multi-million dollar deals. With maturing technologies and business models, AI is progressing towards wide roll-out.

## AI applications abound, from transport to science to health

AI applications are experiencing rapid uptake in a number of sectors where it is possible for them to detect patterns in large volumes of data and model complex, interdependent systems to improve decision making and save costs.

- In the transport sector, autonomous vehicles with virtual driver systems, high-definition maps and optimised traffic routes all promise cost, safety, quality of life and environmental benefits.

- Scientific research uses AI to collect and process large-scale data, to help reproduce experiments and lower their cost, and to accelerate scientific discovery.

- In healthcare, AI systems help diagnose and prevent disease and outbreaks early on, discover treatments and drugs, propose tailored interventions and power self-monitoring tools.

- In criminal justice, AI is used for predictive policing and assessing reoffending risk.

- Digital security applications use AI systems to help automate the detection of and response to threats, increasingly in real time.

- AI applications in agriculture include crop and soil health monitoring and predicting the impact of environmental factors on crop yield.

- Financial services leverage AI to detect fraud, assess credit-worthiness, reduce customer service costs, automate trading and support legal compliance.

- In marketing and advertising, AI mines data on consumer behaviour to target and personalise content, advertising, goods and services, recommendations and prices.

## Trustworthy AI is key to reaping AI's benefits

Alongside benefits, AI raises public policy considerations and efforts are needed to ensure trustworthy, human-centred AI systems. AI – notably some types of ML – raises new types of ethical and fairness concerns. Chief among them are questions of respect for human rights and democratic values, and the dangers of transferring biases from the analogue into the digital world. Some AI systems are so complex that explaining their decisions may be impossible. Designing systems that are transparent about the use of AI and are accountable for their outcomes is critical. AI systems must function properly and in a secure and safe manner.

National policies are needed to promote trustworthy AI systems, including those that encourage investment in responsible AI research and development. In addition to AI technology and computing capacity, AI leverages vast quantities of data. This increases the need for a digital environment that enables access to data, alongside strong data and privacy protections. AI-enabling ecosystems can also support small and medium-sized enterprises as they navigate the AI transition and ensure a competitive environment.

AI will change the nature of work as it replaces and alters components of human labour. Policies will need to facilitate transitions as people move from one job to another, and ensure continuous education, training and skills development.

## AI is a growing policy priority for all stakeholders

In view of the transformative benefits of AI as well as its risks, AI is a growing policy priority for all stakeholders. Many countries have dedicated AI strategies that consider AI as an engine of growth and well-being, seek to educate and recruit the next generation of researchers, and consider how best to address AI challenges. Non-governmental stakeholders – business, technical organisations, academia, civil society and trade unions – and international bodies including the G7, G20, OECD, European Commission and United Nations and are also taking action.

In May 2019 the OECD adopted its Principles on Artificial Intelligence, the first international standards agreed by governments for the responsible stewardship of trustworthy AI, with guidance from a multi-stakeholder expert group.

# 1. The technical landscape

*This chapter characterises the "artificial intelligence (AI) technical landscape", which has evolved significantly from 1950 when Alan Turing first posed the question of whether machines can think. Since 2011, breakthroughs have taken place in the subset of AI called "machine learning", in which machines leverage statistical approaches to learn from historical data and make predictions in new situations. The maturity of machine-learning techniques, along with large datasets and increasing computational power are behind the current expansion of AI. This chapter also provides a high-level understanding of an AI system, which predicts, recommends or decides an outcome to influence the environment. In addition, it details a typical AI system lifecycle from i) design, data and models, including planning and design, data collection and processing and model building and interpretation; ii) verification and validation; iii) deployment; to iv) operation and monitoring. Lastly, this chapter proposes a research taxonomy to support policy makers.*

## A short history of artificial intelligence

In 1950, British mathematician Alan Turing published a paper on computing machinery and intelligence (Turing, 1950[1]) posing the question of whether machines can think. He developed a simple heuristic to test his hypothesis: could a computer have a conversation and answer questions in a way that would trick a suspicious human into thinking the computer was actually a human?[1] The resulting "Turing test" is still used today. That same year, Claude Shannon proposed the creation of a machine that could be taught to play chess (Shannon, 1950[2]). The machine could be trained by using brute force or by evaluating a small set of an opponent's strategic moves (UW, 2006[3]).

Many consider the Dartmouth Summer Research Project in the summer of 1956 as the birthplace of artificial intelligence (AI). At this workshop, the principle of AI was conceptualised by John McCarthy, Alan Newell, Arthur Samuel, Herbert Simon and Marvin Minsky. While AI research has steadily progressed over the past 60 years, the promises of early AI promoters proved to be overly optimistic. This led to an "AI winter" of reduced funding and interest in AI research during the 1970s.

New funding and interest in AI appeared with advances in computation power that became available in the 1990s (UW, 2006[3]). Figure 1.1 provides a timeline of AI's early development.

**Figure 1.1. Timeline of early AI developments (1950s to 2000)**

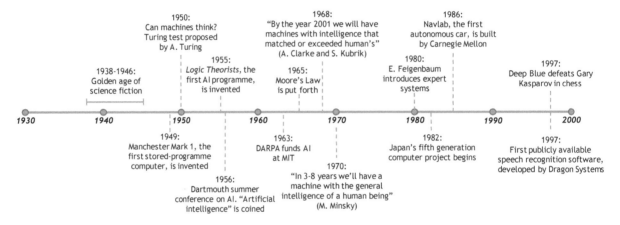

*Source*: Adapted from Anyoha (28 August 2017[4]), "The history of artificial intelligence", http://sitn.hms.harvard.edu/flash/2017/history-artificial-intelligence/.

The AI winter ended in the 1990s as computational power and data storage were advancing to the point that complex tasks were becoming feasible. In 1995, AI took a major step forward with Richard Wallace's development of the Artificial Linguistic Internet Computer Entity that could hold basic conversations. Also in the 1990s, IBM developed a computer named Deep Blue that used a brute force approach to play against world chess champion Gary Kasparov. Deep Blue would look ahead six steps or more and could calculate 330 million positions per second (Somers, 2013[5]). In 1996, Deep Blue lost to Kasparov, but won the rematch a year later.

In 2015, Alphabet's DeepMind launched software to play the ancient game of Go against the best players in the world. It used an artificial neural network that was trained on thousands of human amateur and professional games to learn how to play. In 2016, AlphaGo beat the world's best player at the time, Lee Sedol, four games to one. AlphaGo's developers then let the program play against itself using trial and error, starting from completely random play with a few simple guiding rules. The result was a program (AlphaGo Zero) that trained itself faster and was able to beat the original AlphaGo by 100 games to 0. Entirely from self-play – with no human intervention and using no historical data – AlphaGo Zero surpassed all other versions of AlphaGo in 40 days (Silver et al., 2017[6]) (Figure 1.2).

**Figure 1.2. AlphaGo's rapid self-learning to become best Go player in the world in 40 days**

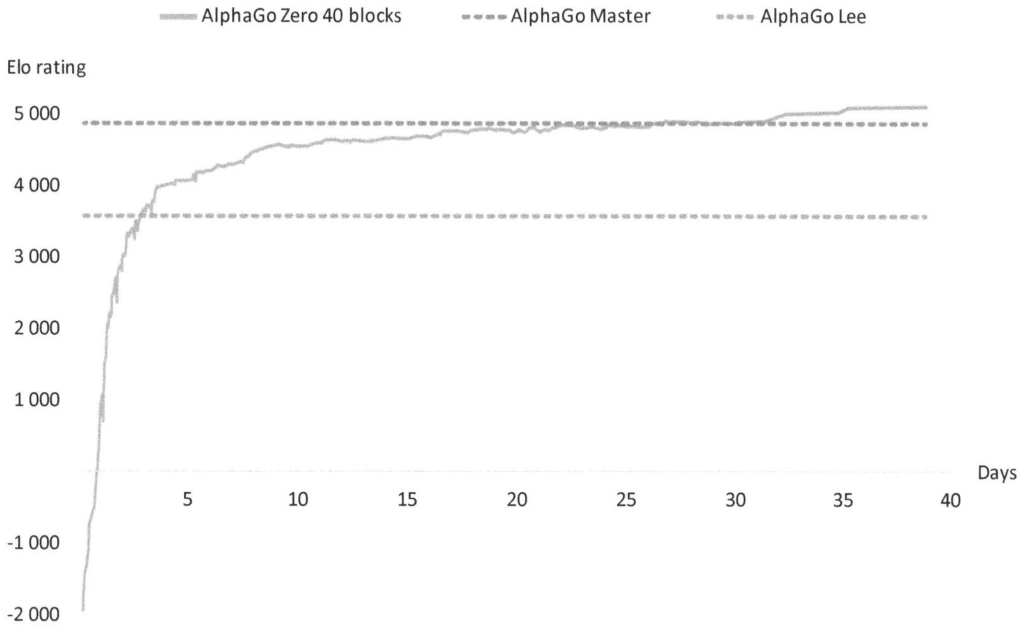

*Source*: adapted from Silver et al. (2017[6]), "Mastering the game of Go without human knowledge", http://dx.doi.org/10.1038/nature24270.

## *Where we are today*

Over the past few years, the availability of big data, cloud computing and the associated computational and storage capacity and breakthroughs in an AI technology called "machine learning" (ML), have dramatically increased the power, availability, growth and impact of AI.

Continuing technological progress is also leading to better and cheaper sensors, which capture more-reliable data for use by AI systems. The amount of data available for AI systems continues to grow as these sensors become smaller and less expensive to deploy. The result is significant progress in many core AI research areas such as:

- natural language processing
- autonomous vehicles and robotics
- computer vision
- language learning.

Some of the most interesting AI developments are outside of computer science in fields such as health, medicine, biology and finance. In many ways, the AI transition resembles the way computers diffused from a few specialised businesses to the broader economy and society in the 1990s. It also recalls how Internet access expanded beyond multinational firms to a majority of the population in many countries in the 2000s. Economies will increasingly need sector "bilinguals". These are people specialised in one area such as economics, biology or law, but also skilled at AI techniques such as ML. The present chapter focuses on applications that are in use or foreseeable in the short and medium term rather than possible longer-term developments such as artificial general intelligence (AGI) (Box 1.1).

---

**Box 1.1. Artificial narrow intelligence versus artificial general intelligence**

Artificial narrow intelligence (ANI) or "applied" AI is designed to accomplish a specific problem-solving or reasoning task. This is the current state-of-the-art. The most advanced AI systems available today, such as Google's AlphaGo, are still "narrow". To some extent, they can generalise pattern recognition such as by transferring knowledge learned in the area of image recognition into speech recognition. However, the human mind is far more versatile.

Applied AI is often contrasted to a (hypothetical) AGI. In AGI, autonomous machines would become capable of general intelligent action. Like humans, they would generalise and abstract learning across different cognitive functions. AGI would have a strong associative memory and be capable of judgment and decision making. It could solve multifaceted problems, learn through reading or experience, create concepts, perceive the world and itself, invent and be creative, react to the unexpected in complex environments and anticipate. With respect to a potential AGI, views vary widely. Experts caution that discussions should be realistic in terms of time scales. They broadly agree that ANI will generate significant new opportunities, risks and challenges. They also agree that the possible advent of an AGI, perhaps sometime during the 21st century, would greatly amplify these consequences.

*Source*: OECD (2017[7]), *OECD Digital Economy Outlook 2017*, http://dx.doi.org/10.1787/9789264276284-en.

---

## What is AI?

There is no universally accepted definition of AI. In November 2018, the AI Group of Experts at the OECD (AIGO) set up a subgroup to develop a description of an AI system. The description aims to be understandable, technically accurate, technology-neutral and applicable to short- and long-term time horizons. It is broad enough to encompass many of the definitions of AI commonly used by the scientific, business and policy communities. As well, it informed the development of the OECD *Recommendation of the Council on Artificial Intelligence* (OECD, 2019[8]).

### *Conceptual view of an AI system*

The present description of an AI system is based on the conceptual view of AI detailed in *Artificial Intelligence: A Modern Approach* (Russel and Norvig, 2009[9]). This view is consistent with a widely used definition of AI as "the study of the computations that make it possible to perceive, reason, and act" (Winston, 1992[10]) and with similar general definitions (Gringsjord and Govindarajulu, 2018[11]).

A conceptual view of AI is first presented as the high-level structure of a generic AI system (also referred to as "intelligent agent") (Figure 1.3). An AI system consists of three main elements: sensors, operational logic and actuators. Sensors collect raw data from the environment, while actuators act to change the state of the environment. The key power of an AI system resides in its operational logic. For a given set of objectives and based on input data from sensors, the operational logic provides output for the actuators. These take the form of recommendations, predictions or decisions that can influence the state of the environment.

**Figure 1.3. A high-level conceptual view of an AI system**

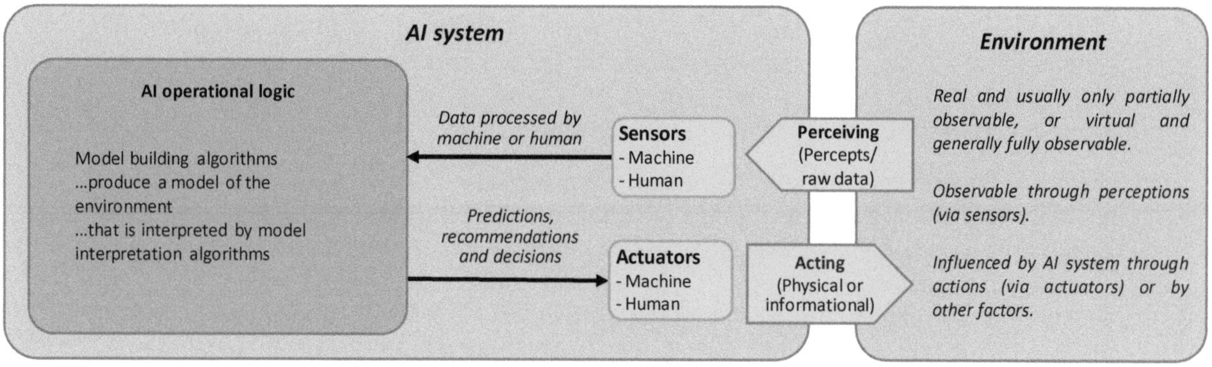

*Source*: As defined and approved by AIGO in February 2019.

A more detailed structure captures the main elements relevant to the policy dimensions of AI systems (Figure 1.4). To cover different types of AI systems and different scenarios, the diagram separates the model building process (such as ML), from the model itself. Model building is also separate from the model interpretation process, which uses the model to make predictions, recommendations and decisions; actuators use these outputs to influence the environment.

**Figure 1.4. Detailed conceptual view of an AI System**

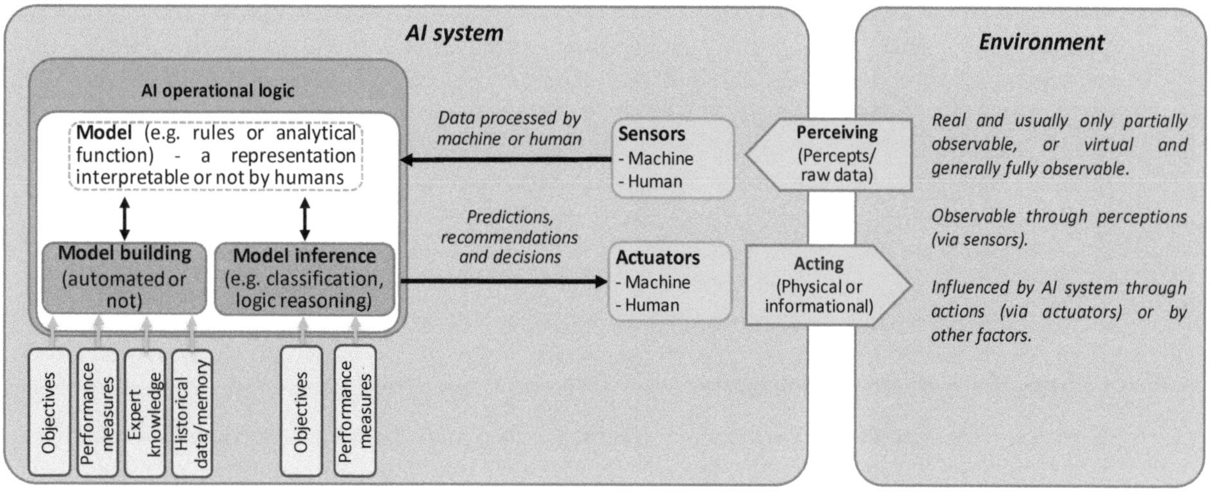

*Source*: As defined and approved by AIGO in February 2019.

## Environment

An environment in relation to an AI system is a space observable through perceptions (via sensors) and influenced through actions (via actuators). Sensors and actuators are either machines or humans. Environments are either real (e.g. physical, social, mental) and usually only partially observable, or else virtual (e.g. board games) and generally fully observable.

## AI system

An AI system is a machine-based system that can, for a given set of human-defined objectives, make predictions, recommendations or decisions influencing real or virtual environments.

It does so by using machine and/or human-based inputs to: i) perceive real and/or virtual environments; ii) abstract such perceptions into models through analysis in an automated manner (e.g. with ML, or manually); and iii) use model inference to formulate options for information or action. AI systems are designed to operate with varying levels of autonomy.

### AI model, model building and model interpretation

The core of an AI system is the AI model, a representation of all or part of the system's external environment that describes the environment's structure and/or dynamics. A model can be based on expert knowledge and/or data, by humans and/or by automated tools (e.g. ML algorithms). Objectives (e.g. output variables) and performance measures (e.g. accuracy, resources for training, representativeness of the dataset) guide the building process. Model inference is the process by which humans and/or automated tools derive an outcome from the model. These take the form of recommendations, predictions or decisions. Objectives and performance measures guide the execution. In some cases (e.g. deterministic rules), a model can offer a single recommendation. In other cases (e.g. probabilistic models), a model can offer a variety of recommendations. These recommendations are associated with different levels of, for instance, performance measures like level of confidence, robustness or risk. In some cases, during the interpretation process, it is possible to explain why specific recommendations are made. In other cases, explanation is almost impossible.

### AI system illustrations

### Credit-scoring system

A credit-scoring system illustrates a machine-based system that influences its environment (whether people are granted a loan). It makes recommendations (a credit score) for a given set of objectives (credit-worthiness). It does so by using both machine-based inputs (historical data on people's profiles and on whether they repaid loans) and human-based inputs (a set of rules). With these two sets of inputs, the system perceives real environments (whether people are repaying loans on an ongoing basis). It abstracts such perceptions into models automatically. A credit-scoring algorithm could, for example, use a statistical model. Finally, it uses model inference (the credit-scoring algorithm) to formulate a recommendation (a credit score) of options for outcomes (providing or denying a loan).

### Assistant for the visually impaired

An assistant for visually impaired people illustrates how a machine-based system influences its environment. It makes recommendations (e.g. how a visually impaired person can avoid an obstacle or cross the street) for a given set of objectives (travel from one place to another). It does so using machine and/or human-based inputs (large tagged image databases of objects, written words and even human faces) for three ends. First, it perceives images of the environment (a camera captures an image of what is in front of a person and sends it to an application). Second, it abstracts such perceptions into models automatically (object recognition algorithms that can recognise a traffic light, a car or an obstacle on the sidewalk). Third, it uses model inference to recommend options for outcomes (providing an audio description of the objects detected in the environment) so the person can decide how to act and thereby influence the environment.

*AlphaGo Zero*

AlphaGo Zero is an AI system that plays the board game Go better than any professional human Go players. The board game's environment is virtual and fully observable. Game positions are constrained by the objectives and the rules of the game. AlphaGo Zero is a system that uses both human-based inputs (the rules of Go) and machine-based inputs (learning based on playing iteratively against itself, starting from completely random play). It abstracts the data into a (stochastic) model of actions ("moves" in the game) trained via so-called reinforcement learning. Finally, it uses the model to propose a new move based on the state of play.

*Autonomous driving system*

Autonomous driving systems illustrate a machine-based system that can influence its environment (whether a car accelerates, decelerates or turns). It makes predictions (whether an object or a sign is an obstacle or an instruction) and/or makes decisions (accelerating, braking, etc.) for a given set of objectives (going from point A to B safely in the least time possible). It does so by using both machine-based inputs (historical driving data) and human-based inputs (a set of driving rules). These inputs are used to create a model of the car and its environment. In this way, it will allow the system to achieve three goals. First, it can perceive real environments (through sensors such as cameras and sonars). Second, it can abstract such perceptions into models automatically (including object recognition; speed and trajectory detection; and location-based data). Third, it can use model inference. For example, the self-driving algorithm can consist of numerous simulations of possible short-term futures for the vehicle and its environment. In this way, it can recommend options for outcomes (to stop or go).

## The AI system lifecycle

In November 2018, AIGO established a subgroup to inform the OECD *Recommendation of the Council on Artificial Intelligence* (OECD, 2019[8]) by detailing the AI system lifecycle. This lifecycle framework does not represent a new standard for the AI lifecycle[2] or propose prescriptive actions. However, it can help contextualise other international initiatives on AI principles.[3]

An AI system incorporates many phases of traditional software development lifecycles and system development lifecycles more generally. However, the AI system lifecycle typically involves four specific phases. The design, data and models phase is a context-dependent sequence encompassing planning and design, data collection and processing, as well as model building and interpretation. This is followed by verification and validation, deployment, and operation and monitoring (Figure 1.5. AI system lifecycle). These phases often take place in an iterative manner and are not necessarily sequential. The decision to retire an AI system from operation may occur at any point during the operation and monitoring phase.

The AI system lifecycle phases can be described as follows:

1. **Design, data and modelling** includes several activities, whose order may vary for different AI systems:

    o **Planning and design** of the AI system involves articulating the system's concept and objectives, underlying assumptions, context and requirements, and potentially building a prototype.

○ **Data collection and processing** includes gathering and cleaning data, performing checks for completeness and quality, and documenting the metadata and characteristics of the dataset. Dataset metadata include information on how a dataset was created, its composition, its intended uses and how it has been maintained over time.

○ **Model building and interpretation** involves the creation or selection of models or algorithms, their calibration and/or training and interpretation.

2. **Verification and validation** involves executing and tuning models, with tests to assess performance across various dimensions and considerations.

3. **Deployment** into live production involves piloting, checking compatibility with legacy systems, ensuring regulatory compliance, managing organisational change and evaluating user experience.

4. **Operation and monitoring** of an AI system involves operating the AI system and continuously assessing its recommendations and impacts (both intended and unintended) in light of objectives and ethical considerations. This phase identifies problems and adjusts by reverting to other phases or, if necessary, retiring an AI system from production.

**Figure 1.5. AI system lifecycle**

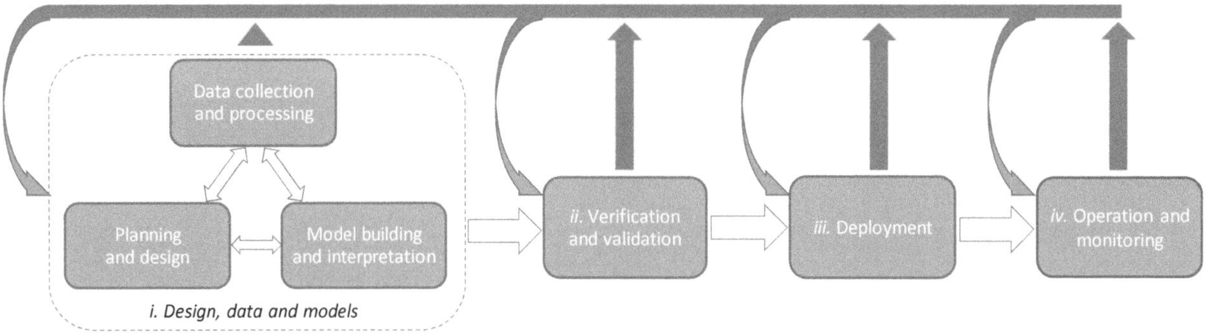

*Source*: As defined and approved by AIGO in February 2019.

The centrality of data and of models that rely on data for their training and evaluation distinguishes the lifecycle of many AI systems from that of more general system development. Some AI systems based on ML can iterate and evolve over time.

## AI research

This section reviews some technical developments with regard to AI research in academia and the private sector that are enabling the AI transition. AI, and particularly its subset called ML, is an active research area in computer science today. A broader range of academic disciplines is leveraging AI techniques for a wide variety of applications.

There is no agreed-upon classification scheme for breaking AI into research streams that is comparable, for example, to the 20 major economics research categories in the *Journal of Economic Literature*'s classification system. This section aims to develop an AI research taxonomy for policy makers to understand some recent AI trends and identify policy issues.

Research has historically distinguished symbolic AI from statistical AI. Symbolic AI uses logical representations to deduce a conclusion from a set of constraints. It requires that researchers build detailed and human-understandable decision structures to translate real-world complexity and help machines arrive at human-like decisions. Symbolic AI is still in widespread use, e.g. for optimisation and planning tools. Statistical AI, whereby machines induce a trend from a set of patterns, has seen increasing uptake recently. A number of applications combine symbolic and statistical approaches. For example, natural language processing (NLP) algorithms often combine statistical approaches (that build on large amounts of data) and symbolic approaches (that consider issues such as grammar rules). Combining models built on both data and human expertise is viewed as promising to help address the limitations of both approaches.

AI systems increasingly use ML. This is a set of techniques to allow machines to learn in an automated manner through patterns and inferences rather than through explicit instructions from a human. ML approaches often teach machines to reach an outcome by showing them many examples of correct outcomes. However, they can also define a set of rules and let the machine learn by trial and error. ML is usually used in building or adjusting a model, but can also be used to interpret a model's results (Figure 1.6). ML contains numerous techniques that have been used by economists, researchers and technologists for decades. These range from linear and logistic regressions, decision trees and principle component analysis to deep neural networks.

**Figure 1.6. The relationship between AI and ML**

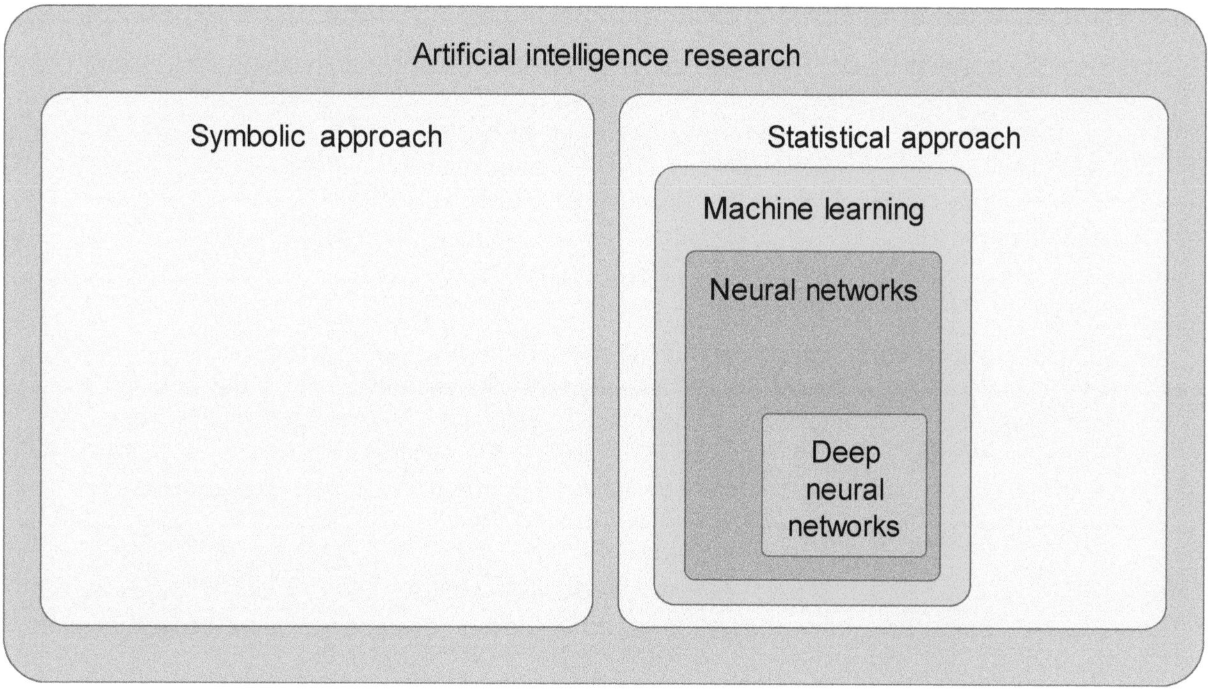

*Source*: Provided by the Massachusetts Institute of Technology (MIT)'s Internet Policy Research Initiative (IPRI).

In economics, regression models use input data to make predictions in such a way that researchers can interpret the coefficients (weights) on the input variables, often for policy reasons. With ML, people may not be able to understand the models themselves. Additionally, ML problems tend to work with many more variables than is common in economics. These variables, known as "features", typically number in the thousands or higher. Larger data sets can range from tens of thousands to hundreds of millions of observations. At this scale, researchers rely on more sophisticated and less-understood techniques such as neural networks to make predictions. Interestingly, one core research area of ML is trying to reintroduce the type of explainability used by economists in these large-scale models (see Cluster 4 below).

The real technology behind the current wave of ML applications is a sophisticated statistical modelling technique called "neural networks". This technique is accompanied by growing computational power and the availability of massive datasets ("big data"). Neural networks involve repeatedly interconnecting thousands or millions of simple transformations into a larger statistical machine that can learn sophisticated relationships between inputs and outputs. In other words, neural networks modify their own code to find and optimise links between inputs and outputs. Finally, deep learning is a phrase that refers to particularly large neural networks; there is no defined threshold as to when a neural net becomes "deep".

This evolving dynamic in AI research is paired with continual advances in computational abilities, data availability and neural network design. Together, they mean the statistical approach to AI will likely continue as an important part of AI research in the short term. As a result, policy makers should focus their attention on AI developments that will likely have the largest impact over the coming years and represent some of the most difficult policy challenges. These challenges include unpacking the machines' decisions and making the decision-making process more transparent. Policy makers should also keep in mind that most dynamic AI approaches – statistical AI, specifically "neural networks" – are not relevant for all types of problems. Other AI approaches, and coupling symbolic and statistical methods, remain important.

There is no widely agreed-upon taxonomy for AI research or for the subset of ML. The taxonomy proposed in the next subsection represents 25 AI research streams. They are organised into four broad categories and nine sub-categories, mainly focused on ML. In traditional economic research traditions, researchers may focus on a narrow research area. AI researchers commonly work across multiple clusters simultaneously to solve open research problems.

### Cluster 1: ML applications

The first broad research category applies ML methods to solve various practical challenges in the economy and society. Examples of applied ML are emerging in much the same way as Internet connectivity transformed certain industries first and then swept across the entire economy. Chapter 3 provides a range of examples of AI applications emerging across OECD countries. The research streams in Table 1.1 represent the largest areas of research linked to real-world application development.

**Table 1.1. Cluster 1: Application areas**

| | | |
|---|---|---|
| Application areas | Using ML | Natural language processing |
| | | Computer vision |
| | | Robotic navigation |
| | | Language learning |
| | Contextualising ML | Algorithmic game theory and computational social choice |
| | | Collaborative systems |

*Source*: Provided by the MIT's IPRI.

Core applied research areas that use ML include natural language processing, computer vision and robotic navigation. Each of these three research areas represents a rich and expanding research field. Research challenges can be confined to just one area or can span multiple streams. For example, researchers in the United States are combining NLP of free text mammogram and pathology notes with computer vision of mammograms to aid with breast cancer screening (Yala et al., 2017[12]).

Two research lines focus on ways to contextualise ML. Algorithmic game theory lies at the intersection of economics, game theory and computer science. It uses algorithms to analyse and optimise multi-period games. Collaborative systems are an approach to large challenges where multiple ML systems combine to tackle different parts of complex problems.

## Cluster 1: Policy relevance

Several relevant policy issues are linked to AI applications. These include the future of work, the potential impact of AI, and human capital and skills development. They also include understanding in which situations AI applications may or may not be appropriate in sensitive contexts. Other relevant issues include AI's impact on industry players and dynamics, government open data policies, regulations for robotic navigation and privacy policies that govern the collection and use of data.

## Cluster 2: ML techniques

The second broad category of research focuses on the techniques and paradigms used in ML. Similar to quantitative methods research in the social sciences, this line of research builds and supplies the technical tools and approaches used in machine-learning applications (Table 1.2).

**Table 1.2. Cluster 2: ML techniques**

| ML techniques | Techniques | Deep learning |
| | | Simulation-based learning |
| | | Crowdsourcing and human computation |
| | | Evolutionary computing |
| | | Techniques beyond neural networks |
| | Paradigms | Supervised learning |
| | | Reinforcement learning |
| | | Generative models/generative adversarial networks |

*Source*: Provided by the MIT's IPRI.

The category is dominated by neural networks (of which "deep learning" is a subcategory) and forms the basis for most ML today. ML techniques also include various paradigms for helping the system learn. Reinforcement learning trains the system in a way that mimics the way humans learn via trial and error. The algorithms are not provided explicit tasks, but rather learn by trying different options in rapid succession. Based on rewards or punishments as outcomes, they adapt accordingly. This has been referred to as relentless experimentation (Knight, 2017[13]).

Generative models, including generative adversarial networks, train a system to produce new data similar to an existing dataset. They are an exciting area of AI research because they pit two or more unsupervised neural networks against each other in a zero-sum game. In game theory terms, they function and learn as a set of rapidly repeated games. By setting the systems against each other at computationally high speeds, the systems can learn profitable strategies. This is particularly the case in structured environments with clear rules, such as the game of Go with AlphaGo Zero.

## Cluster 2: Policy relevance

Several public policy issues are relevant to the development and deployment of ML technologies. These issues include supporting better training data sets; funding for academic research and basic science; policies to create "bilinguals" who can combine AI skills with other competencies; and computing education. For example, research funding from the Canadian government supported breakthroughs that led to the extraordinary success of modern neural networks (Allen, 2015[14]).

## Cluster 3: Ways of improving ML/optimisations

The third broad category of research focuses on ways to improve and optimise ML tools. It breaks down research streams based on the time horizon for results (current, emerging and future) (Table 1.3). Short-term research is focusing on speeding up the deep-learning process. It does this either via better data collection or by using distributed computer systems to train the algorithm.

**Table 1.3. Cluster 3: Ways of improving ML/optimisations**

| | | |
|---|---|---|
| Ways of improving ML | Enabling factors (current) | Faster deep learning |
| | | Better data collection |
| | | Distributed training algorithms |
| | Enabling factors (emerging) | Performance on low-power devices |
| | | Learning to learn/meta learning |
| | | AI developer tools |
| | Enabling factors (future) | Understanding neural networks |
| | | One-shot learning |

*Source*: Provided by the MIT's IPRI.

Researchers are also focused on enabling ML on low-power devices such as mobile phones and other connected devices. Significant progress has been made on this front. Projects such as Google's Teachable Machine now offer open-source ML tools light enough to run in a browser (Box 1.2). Teachable Machine is just one example of emerging AI development tools meant to expand the reach and efficiency of ML. There are also significant advances in the development of dedicated AI chips for mobile devices.

ML research with a longer time horizon includes studying the mechanisms that allow neural networks to learn so effectively. Although neural networks have proven to be a powerful ML technique, understanding of how they operate is still limited. Understanding these processes would make it possible to engineer neural networks on a deeper level. Longer-term research is also looking at ways to train neural networks using much smaller sets of training data, sometimes referred to as "one-shot learning". It is also generally trying to make the training process more efficient. Large models can take weeks or months to train and require hundreds of millions of training examples.

**Box 1.2. Teachable Machine**

Teachable Machine is a Google experiment that allows people to train a machine to detect different scenarios using a camera built into a phone or computer. The user takes a series of pictures for three different scenarios (e.g. different facial expressions) to train the teachable machine. The machine then analyses the photos in the training data set and can use them to detect different scenarios. For example, the machine can play a sound every time the person smiles in a camera range. Teachable Machine stands out as an ML project because the neural network runs exclusively in the user's browser without any need for outside computation or data storage (Figure 1.7).

**Figure 1.7. Training a machine using a computer's camera**

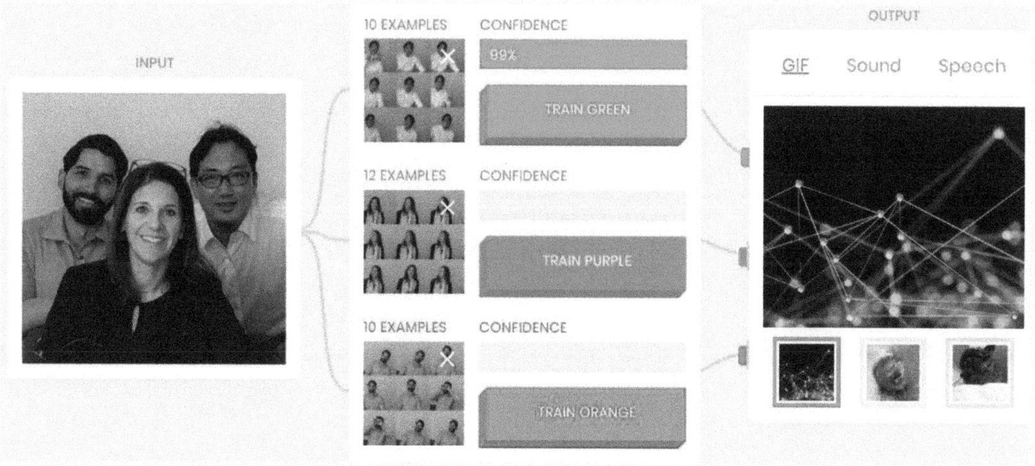

*Source*: https://experiments.withgoogle.com/ai/teachable-machine.

## Cluster 3: Policy relevance

The policy relevance of the third cluster includes the implications of running ML on stand-alone devices and thus of not necessarily sharing data on the cloud. It also includes the potential to reduce energy use, and the need to develop better AI tools to expand its beneficial uses.

## Cluster 4: Considering the societal context

The fourth broad research category examines the context for ML from technical, legal and social perspectives. ML systems increasingly rely on algorithms to make important decisions. Therefore, it is important to understand how bias can be introduced, how bias can propagate and how to eliminate bias from outcomes. One of the most active research areas in ML is concerned with transparency and accountability of AI systems (Table 1.4). Statistical approaches to AI have led to less human-comprehensible computation in algorithmic decisions. These can have significant impacts on the lives of individuals – from bank loans to parole decisions (Angwin et al., 2016[15]). Another category of contextual ML research involves steps to ensure the safety and integrity of these systems. Researchers' understanding of how neural networks arrive at decisions is still at an early stage. Neural networks can often be tricked using simple methods such as changing a few pixels in a picture (Ilyas et al., 2018[16]). Research in these streams seeks to defend systems against inadvertent introduction of unintended information and adversarial attacks. It also aims to verify the integrity of ML systems.

## Cluster 4: Policy relevance

Several relevant policy issues are linked to the context surrounding ML. These include requirements for algorithmic accountability, combating bias, the impact of ML systems, product safety, liability and security (OECD, 2019[8]).

**Table 1.4. Cluster 4: Refining ML with context**

| | | |
|---|---|---|
| Refining ML with context | Explainability | Transparency and accountability |
| | | Explaining individual decisions |
| | | Simplification into human-comprehensible algorithms |
| | | Fairness/bias |
| | | Debug-ability |
| | Safety and reliability | Adversarial examples |
| | | Verification |
| | | Other classes of attacks |

*Source*: Provided by the MIT's IPRI.

# References

Allen, K. (2015), "How a Toronto professor's research revolutionized artificial intelligence.", *The Star*, 17 April, https://www.thestar.com/news/world/2015/04/17/how-a-toronto-professors-research-revolutionized-artificial-intelligence.html.  [14]

Angwin, J. et al. (2016), "Machine bias: There's software used across the country to predict future criminals. And it's biased against blacks", *ProPublica*, https://www.propublica.org/article/machine-bias-risk-assessments-in-criminal-sentencing.  [15]

Anyoha, R. (28 August 2017), "The history of artificial intelligence", Harvard University Graduate School of Arts and Sciences blog, http://sitn.hms.harvard.edu/flash/2017/history-artificial-intelligence/.  [4]

Gringsjord, S. and N. Govindarajulu (2018), *Artificial Intelligence*, The Stanford Encyclopedia of Philosophy Archive, https://plato.stanford.edu/archives/fall2018/entries/artificial-intelligence/.  [11]

Ilyas, A. et al. (2018), *Blackbox Adversarial Attacks with Limited Queries and Information*, presentation at the 35th International Conference on Machine Learning, Stockholmsmässan Stockholm, 10-15 July, pp. 2142-2151.  [16]

Knight, W. (2017), "5 big predictions for artificial intelligence in 2017", *MIT Technology Review*, 4 January, https://www.technologyreview.com/s/603216/5-big-predictions-for-artificial-intelligence-in-2017/.  [13]

OECD (2019), *Recommendation of the Council on Artificial Intelligence*, OECD, Paris.  [8]

OECD (2017), *OECD Digital Economy Outlook 2017*, OECD Publishing, Paris, http://dx.doi.org/10.1787/9789264276284-en.  [7]

Russel, S. and P. Norvig (2009), *Artificial Intelligence: A Modern Approach, 3rd edition*, Pearson, London, http://aima.cs.berkeley.edu/.  [9]

Shannon, C. (1950), "XXII. Programming a computer for playing chess", *The London, Edinburgh and Dublin Philosophical Magazine and Journal of Science*, Vol. 41/314, pp. 256-275.  [2]

Silver, D. et al. (2017), "Mastering the game of Go without human knowledge", *Nature*, Vol. 550/7676, pp. 354-359, http://dx.doi.org/10.1038/nature24270.  [6]

Somers, J. (2013), "The man who would teach machines to think", *The Atlantic*, November, https://www.theatlantic.com/magazine/archive/2013/11/the-man-who-would-teach-machines-to-think/309529/.  [5]

Turing, A. (1950), "Computing machinery and intelligence", in *Parsing the Turing Test*, Springer, Dordrecht, pp. 23-65.  [1]

UW (2006), *History of AI*, University of Washington, History of Computing Course (CSEP 590A), https://courses.cs.washington.edu/courses/csep590/06au/projects/history-ai.pdf.  [3]

Winston, P. (1992), *Artificial Intelligence*, Addison-Wesley, Reading, MA, https://courses.csail.mit.edu/6.034f/ai3/rest.pdf.  [10]

Yala, A. et al. (2017), "Using machine learning to parse breast pathology reports", *Breast Cancer Research and Treatment*, Vol. 161/2, pp. 201-211.  [12]

Notes

[1] These tests were conducted using typed or relayed messages rather than voice.

[2] Work on the System Development Lifecycle has been conducted, among others, by the National Institute of Standards. More recently, standards organisations such as the International Organization for Standardization (ISO) SC 42 have begun to explore the AI lifecycle.

[3] The Institute for Electrical and Electronics Engineers' Global Initiative on Ethics of Autonomous and Intelligent Systems is an example.

# 2. The economic landscape

*This chapter describes the economic characteristics of artificial intelligence (AI) as an emerging general-purpose technology with the potential to lower the cost of prediction and enable better decisions. Through less expensive and more accurate predictions, recommendations or decisions, AI promises to generate productivity gains, improve well-being and help address complex challenges. Speed of adoption varies across companies and industries, since leveraging AI requires complementary investments in data, skills, digitalisation of workflows and the capacity to adapt organisational processes. Additionally, AI has been a growing target area for investment and business development. Private equity investment in AI start-ups has accelerated since 2016, doubling from 2016 to 2017 to reach USD 16 billion. AI start-ups attracted 12% of worldwide private equity investments in the first half of 2018, a significant increase from just 3% in 2011. Investment in AI technologies is expected to continue its upward trend as these technologies mature.*

The statistical data for Israel are supplied by and under the responsibility of the relevant Israeli authorities. The use of such data by the OECD is without prejudice to the status of the Golan Heights, East Jerusalem and Israeli settlements in the West Bank under the terms of international law.

Economic characteristics of artificial intelligence

### *Artificial intelligence enables more readily available prediction*

From an economic point of view, recent advances in artificial intelligence (AI) either decrease the cost of prediction or improve the quality of predictions available at the same cost. Many aspects of decision making are separate from prediction. However, improved, inexpensive and widely accessible AI prediction could be transformative because prediction is an input into much of human activity.

As the cost of AI prediction has decreased, more opportunities to use prediction have emerged, as with computers in the past. The first AI applications were long-recognised as prediction problems. For example, machine learning (ML) predicts loan defaults and insurance risk. As their cost decreases, some human activities are being reframed as prediction issues. In medical diagnosis, for example, a doctor uses data about a patient's symptoms and fills in missing information about the cause of those symptoms. The process of using data to complete missing information is a prediction. Object classification is also a prediction issue: a human's eyes take in data in the form of light signals and the brain fills in the missing information of a label.

AI, through less expensive prediction, has a large number of applications because prediction is a key input into decision making. In other words, prediction helps make decisions, and decision making is everywhere. Managers make important decisions around hiring, investments and strategy, and less important decisions around which meetings to attend and what to say during these meetings. Judges make important decisions about guilt or innocence, procedures and sentencing, and smaller decisions about a specific paragraph or motion. Similarly, individuals make decisions constantly – from whether to marry to what to eat or what song to play. A key challenge in decision making is dealing with uncertainty. Because prediction reduces uncertainty, it is an input into all these decisions and can lead to new opportunities.

### *Machine prediction is a substitute for human prediction*

Another relevant economic concept is substitution. When the price of a commodity (such as coffee) falls, not only do people buy more of it, they also buy fewer substitute products (such as tea). Thus, as machine prediction becomes less expensive, machines will substitute for humans in prediction tasks. This means that reduction of labour related to prediction will be a key impact of AI on human work.

Just as computers meant that few people now perform arithmetic as part of their work, AI will mean that fewer people will have prediction tasks. For example, transcription – conversion of spoken words into text – is prediction in that it fills in missing information on the set of symbols that match the spoken words. AI is already quicker and more accurate than many humans whose work involves transcription.

### *Data, action and judgment complement machine prediction*

As the price of a commodity (e.g. coffee) falls, people buy more of its complements (e.g. cream and sugar). Identifying the complements to prediction, then, is a key challenge with respect to recent advances in AI. While prediction is a key input into decision making, a prediction is not a decision in itself. The other aspects of a decision are complements to AI: data, action and judgment.

**Data** is the information that goes into a prediction. Many recent developments in AI depend on large quantities of digital data for AI systems to predict based on past examples. In general, the more past examples, the more accurate the predictions. Thus, access to large quantities of data is a more valuable asset to organisations because of AI. The strategic value of data is subtle since it depends on whether the data are useful to predict something important to an organisation. Value also depends on whether the data are only available historically or whether an organisation can collect continued feedback over time. The ability to continue to learn through new data can generate sustained competitive advantage (Agrawal, Gans and Goldfarb, 2018[1]).

More new tasks come from the other elements of a decision: **action** and **judgment**. Some *actions* are inherently more valuable when performed by a human rather than a machine (e.g. actions by professional athletes, child carers or salespeople). Perhaps most important is the concept of *judgment*: the process of determining the reward to a particular action in a particular environment. When AI is used for predictions, a human must decide what to predict and what to do with the predictions.

### *Implementing AI in organisations requires complementary investments and process changes*

Like computing, electrification and the steam engine, AI can be seen as a general-purpose technology (Bresnahan and Trajtenberg, 1992[2]; Brynjolfsson, Rock and Syverson, 2017[3]). This means it has potential to substantially increase productivity in a wider variety of sectors. At the same time, the effect of AI requires investment in a number of complementary inputs. It may lead an organisation to change its overall strategy.

In the AI context, organisations need to make a number of complementary investments before AI has a significant impact on productivity. These investments involve infrastructure for the continued collection of data, specialised workers that know how to use data, and changes in processes that take advantage of new opportunities arising from reduced uncertainty.

Many processes in every organisation exist to make the best of a situation in the face of uncertainty rather than to serve customers in the best way possible. Airport lounges, for example, make customers comfortable while they wait for their plane. If passengers had accurate predictions of how long it would take to get to the airport and through security, lounges might not be needed.

The scope of opportunities offered by better predictions is expected to vary across companies and industries. Google, Baidu and other large digital platform companies are well-positioned to benefit from major investments in AI. On the supply side, they already have systems in place to collect data. On the demand side, having enough customers to justify the high fixed costs of investment in the technology is in its early stages. Many other businesses have not fully digitised their workflows, and cannot yet apply AI tools directly into existing processes. As costs fall over time, however, these businesses will recognise the opportunities that are possible by reducing uncertainty. Driven by their needs, they will follow industry leaders and invest in AI.

### Private equity investments in AI start-ups

AI investment as a whole is growing fast and AI already has significant business impact. MGI (2017[4]) estimated that USD 26 billion to USD 39 billion had been invested in AI worldwide in 2016. Of this amount, internal corporate investment represented about 70%, investment in AI start-ups some 20% and AI acquisitions represented some 10% (Dilda, 2017[5]). Large technology companies made three-quarters of these investments. Outside the technology sector, AI adoption is at an early stage; few firms have deployed

AI solutions at scale. Large companies in other digitally mature sectors with data to leverage, notably in the financial and automotive sectors, are also adopting AI.

Large technology companies are acquiring AI start-ups at a rapid pace. According to CBI (2018[6]), the companies that have acquired the most AI start-ups since 2010 include Google, Apple, Baidu, Facebook, Amazon, Intel, Microsoft, Twitter and Salesforce. Several AI cybersecurity start-ups were acquired in 2017 and early 2018. For example, Amazon and Oracle purchased Sqrrl and Zenedge, respectively.

AI start-ups are also acquisition targets for companies in more traditional industries. These include, notably, automotive companies; healthcare companies such as Roche Holding or Athena Health; and insurance and retail companies.

After five years of steady increases, private equity investment in AI start-ups has accelerated since 2016. The amount of private equity invested doubled between 2016 and 2017 (Figure 2.1). It is estimated that more than USD 50 billion was invested in AI start-ups between 2011 and mid-2018 (Box 2.1).

**Figure 2.1. Total estimated investments in AI start-ups, 2011-17 and first semester 2018**

By start-up location

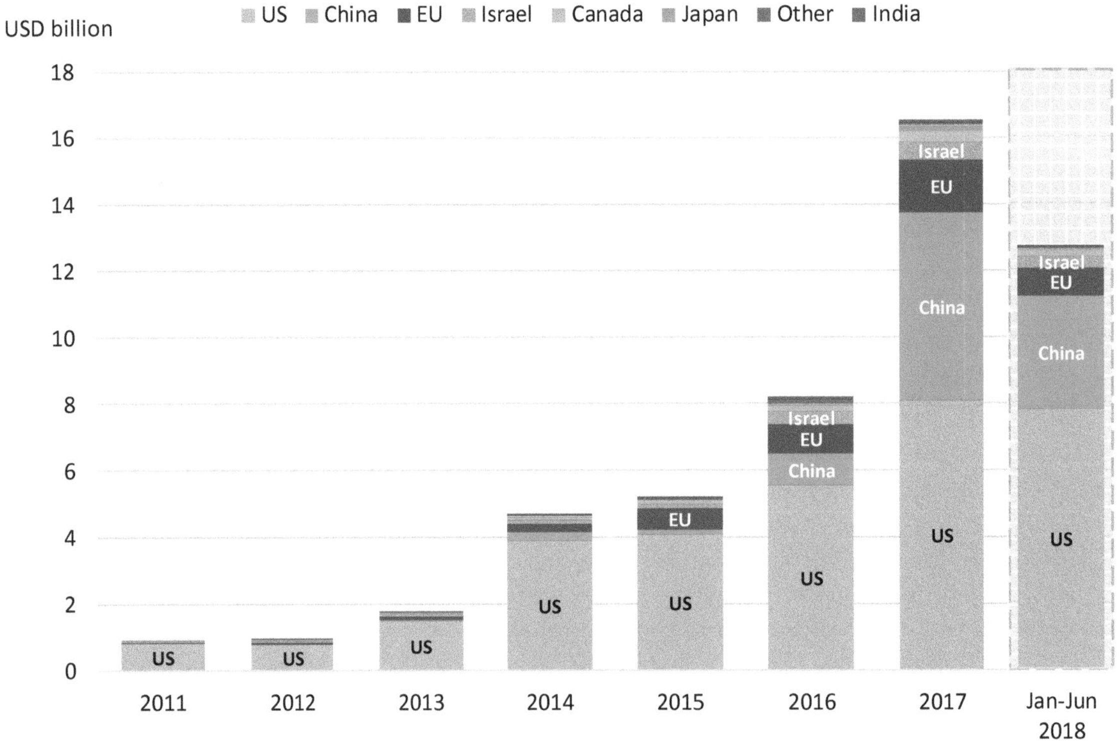

*Note*: Estimates for 2018 may be conservative, as they do not account for a likely lag in reporting (see Box 2.1. Methodological note).
*Source*: OECD estimation, based on Crunchbase (July 2018), www.crunchbase.com.

**Box 2.1. Methodological note**

This section estimates private equity investments in AI start-ups based on Crunchbase (July 2018 version). Crunchbase is a commercial database on innovative companies created in 2007 that contains information on more than 500 000 entities in 199 countries. Breschi, Lassébie and Menon (2018[7]) benchmark Crunchbase with other aggregate data sources. They find consistent patterns for a broad range of countries, including most OECD countries (with the exception of Japan and Korea). Consistent patterns were also found for Brazil, the Russian Federation, India, the People's Republic of China (hereafter "China") and South Africa. Companies in Crunchbase are classified into one or several technological areas, taken from a list of 45 groups.

Caveats to using Crunchbase include the broadly defined scope of the database, the reliability of self-reported information and sample selection issues. In particular, input of new deals into the database likely takes time and the delay may vary across countries. Start-ups may also increasingly self-categorise as AI start-ups because of investors' growing interest in AI.

In this report, AI start-ups correspond to those companies founded after 2000 and categorised in the "artificial intelligence" technological area of Crunchbase (2 436 companies). They also include companies that used AI keywords in their short description of their activities (an additional 689 companies). Three types of keywords are considered to be AI-related. The first type is generic AI keywords, notably "artificial intelligence", "AI", "machine learning" and "machine intelligence". The second type of keywords pertains to AI techniques, notably "neural network", "deep learning" and "reinforcement learning". The third type refers to fields of AI applications, notably "computer vision", "predictive analytics", "natural language processing", "autonomous vehicles", "intelligent systems" and "virtual assistant".

More than one-quarter (26%) of investment deals in AI start-ups included in the database do not report investments by venture capitalists. This analysis estimates the amounts of these deals by using the average amount invested in smaller deals (considering only deals of less than USD 10 million) for the same period and the same country. The rationale for excluding larger deals is that their amounts are more likely to be public information. The estimated value of non-disclosed deals represents about 6% of the total value from 2011 to mid-2018, which may be conservative. The numbers for the first half of 2018 are likely to be conservative because reporting is often not immediate.

## *AI now represents over 12% of private equity investments in start-ups*

AI start-ups attracted around 12% of all worldwide private equity investments in the first half of 2018, a steep increase from just 3% in 2011 (Figure 2.2). All countries analysed increased their share of investments in start-ups focused on AI. About 13% of investments in start-ups in the United States and China were in AI start-ups in the first half of 2018. Most dramatically, Israel has seen the share of investments in AI start-ups jump from 5% to 25% between 2011 and the first half of 2018; autonomous vehicles (AVs) represented 50% of the investments in 2017.

**Figure 2.2. AI as a share of private equity investments in start-ups, 2011 to 2017
and first semester 2018**

Percentage of total investment deals

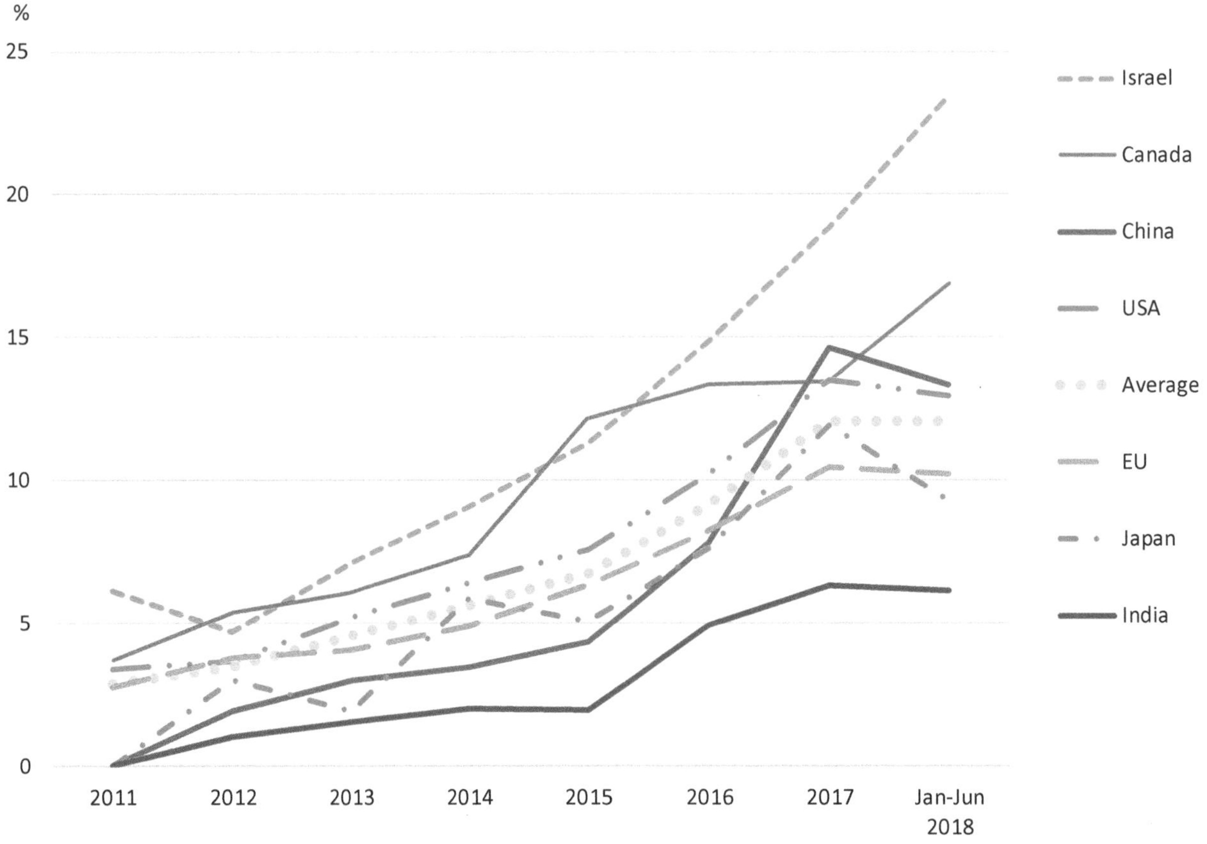

*Note*: The percentage for 2018 only covers the first half of the year (see Box 2.1. Methodological note).
*Source*: OECD estimation, based on Crunchbase (July 2018), www.crunchbase.com.

## *The United States and China account for most AI start-up investments*

Start-ups operating in the United States account for most AI start-up equity investments worldwide. This is true for the number of investment transactions ("deals") and amounts invested, which represents two-thirds of the total value invested since 2011 (Figure 2.1). These facts are unsurprising, considering that the United States accounts for 70-80% of global venture capital investments across all technologies (Breschi, Lassébie and Menon, 2018[7]).

China has seen a dramatic upsurge in AI start-up investment since 2016. It now appears to be the second player globally in terms of the value of AI equity investments received. From just 3% in 2015, Chinese companies attracted 36% of global AI private equity investment in 2017. They maintained an average of 21% from 2011 through to mid-2018.

The European Union accounted for 8% of global AI equity investment in 2017. This represents an important increase for the region as a whole, which accounted for just 1% of this investment in 2013. However, member states varied widely in terms of investment levels. Start-ups in the United Kingdom received 55% of the European Union total investment between 2011 and mid-2018, followed by German (14%) and French ventures (13%). This means the remaining 25 countries shared less than 20% of all private AI equity investments received in the European Union (Figure 2.3).

**Figure 2.3. Private equity investments in AI start-ups based in the European Union, 2011 to mid-2018**

Percentage of total amount invested in EU-based start-ups over period

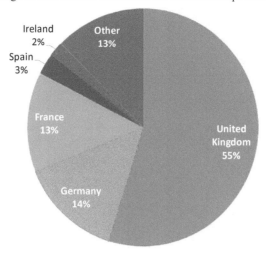

*Note*: The percentage for 2018 only covers the first half of the year.
*Source*: OECD estimation, based on Crunchbase (July 2018), www.crunchbase.com.

Together, the United States, China and the European Union represent over 93% of total AI private equity investment from 2011 to mid-2018. Beyond these leaders, start-ups in Israel (3%) and Canada (1.6%) also played a role.

### *The volume of AI deals grew until 2017, but so did their size*

The number of investment transactions grew globally, from fewer than 200 investment deals to more than 1 400 over 2011-17. This represents a 35% compound annual growth rate from 2011 to the first half of 2018 (Figure 2.4). Start-ups based in the United States attracted a significant portion of all investment deals, rising from 130 to about 800 over 2011-17. The European Union has also seen an increase in the number of deals, from about 30 to about 350 during the same period.

China-based start-ups signed fewer deals than companies in the United States or the European Union, going from none to about 60 over 2011-17. However, the high total value of investment in China implies the average value of these deals was considerably higher than in the European Union.

The large average size of investments in China is in line with a general trend of deals with an increase in per-investment value. In 2012 and 2013, close to nine out of ten reported investment deals were worth less than USD 10 million. Only one out of ten deals was worth between USD 10-100 million. No deals were worth more than USD 100 million. By 2017, more than two deals out of ten were larger than USD 10 million and close to 3% were larger than USD 100 million. The trend accentuated in the first half of 2018, with 40% of reported deals worth more than USD 10 million and 4.4% worth over USD 100 million.

In terms of value, "mega-deals" (those larger than USD 100 million) represented 66% of the total amount invested in AI start-ups in the first half of 2018. These figures reflect the maturing of AI technologies and investor strategies, with larger investments focused on fewer AI companies. For example, the Chinese start-up Toutiao attracted the largest investment in 2017 (USD 3 billion). The company is an AI-powered content recommendation system based on data mining that suggests relevant, personalised information to users in China through social network analysis.

**Figure 2.4. Number of private equity investments in AI start-ups, by start-up location**

2011-17 and first semester 2018

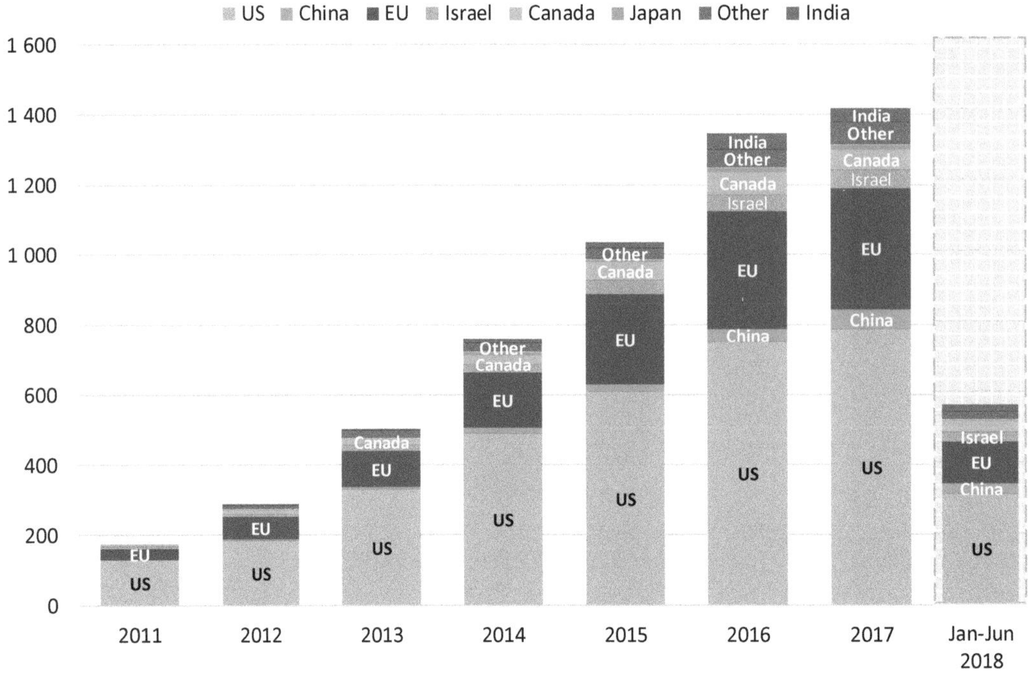

*Note*: Estimates for 2018 may be conservative, as they do not account for a likely lag in reporting (see Box 2.1. Methodological note).
*Source*: OECD estimation, based on Crunchbase (July 2018), www.crunchbase.com.

Since 2016, Israel (Voyager Labs), Switzerland (Mindmaze), Canada (LeddarTech and Element AI) and the United Kingdom (Oaknorth and Benevolent AI) have all seen deals worth USD 100 million or more. This highlights dynamic AI activity beyond the United States and China.

## *Investment patterns vary across countries and regions*

The total amount invested and the global number of deals have increased greatly since 2011, but with wide variations in investment profiles between countries and regions.

In particular, the profile of investments in Chinese start-ups appears different from that of the rest of the world. Individual private equity investments in Chinese AI start-ups registered in Crunchbase were worth an average of USD 150 million in 2017 and in the first half of 2018. By comparison, the average investment size in 2017 in other countries was just one-tenth of that amount.

Overall, three patterns can be observed. First, there are few Chinese start-ups, but they have large investments. Second, EU start-ups have a steadily increasing number of smaller investments. The average per investment increased from USD 3.2 million in 2016 to USD 5.5 million in 2017 to USD 8.5 million in the first half of 2018. Third, the United States has a steadily increasing number of larger investments. The average per investment increased from USD 9.5 million in 2016 to USD 13.2 million in 2017 to USD 32 million in the first half of 2018. These differences in investment profiles remain notable even when deals over USD 100 million are excluded from the sample (Table 2.1 and Table 2.2).

The investment profiles described above are not limited to AI start-ups. They are, instead, true across industries. In 2017, Chinese start-ups across all industries raised USD 200 million on average per investment round. Meanwhile, start-ups in the United States and the European Union raised on average USD 22 million and USD 10 million, respectively.

**Figure 2.5. Size of investment transactions, 2012-17 and first semester 2018**

Percentage of total number of investment deals

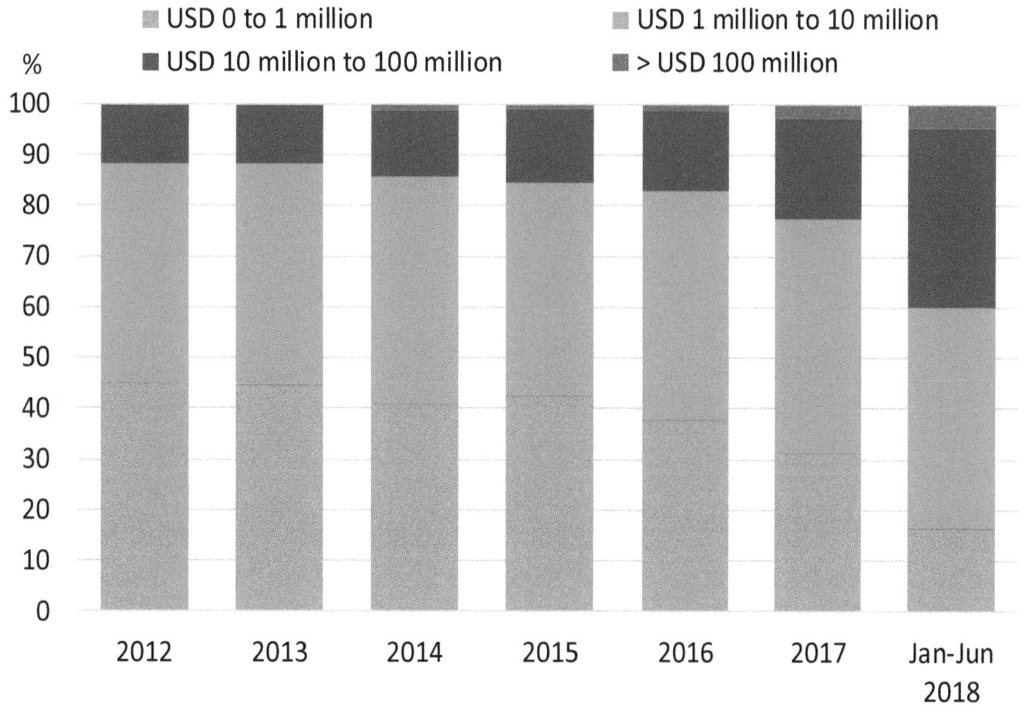

*Note*: The percentages for 2018 only cover the first half of the year.
*Source*: OECD estimation, based on Crunchbase (July 2018), www.crunchbase.com.

**Table 2.1. Average amount raised per deal, for deals up to USD 100 million**

USD million

|  | Canada | China | EU | Israel | Japan | United States |
|---|---|---|---|---|---|---|
| **2015** | 2 | 12 | 2 | 4 | 4 | 6 |
| **2016** | 4 | 20 | 3 | 6 | 5 | 6 |
| **2017** | 2 | 26 | 4 | 12 | 14 | 8 |

*Source*: OECD estimation, based on Crunchbase (April 2018), www.crunchbase.com.

**Table 2.2. Average amount raised per deal, for all AI deals**

USD million

|  | Canada | China | EU | Israel | Japan | United States |
|---|---|---|---|---|---|---|
| **2015** | 2 | 12 | 3 | 4 | 4 | 8 |
| **2016** | 4 | 73 | 3 | 6 | 5 | 10 |
| **2017** | 8 | 147 | 6 | 12 | 14 | 14 |

*Source*: OECD estimation, based on Crunchbase (April 2018), www.crunchbase.com.

*Autonomous vehicle start-ups are receiving significant funding*

Levels of private equity investment in AI vary widely by field of application. AVs represent an increasing share of private equity investments in AI start-ups. Until 2015, AVs represented less than 5% of total investments in AI start-ups. By 2017, AVs represented 23% of the total, growing to 30% by mid-2018. The bulk of venture capital investment in AV start-ups went to US-based start-ups (80% between 2017 and mid-2018). This was followed by AV start-ups based in China (15%), Israel (3%) and the European Union (2%). The growth is due to a dramatic increase in the per-investment amount; the actual number of investments remained fairly constant (87 in 2016 and 95 in 2017). In the United States, the average amount per investment in this sector increased ten-fold from USD 20 million to close to USD 200 million between 2016 and the first half of 2018. This was in large part due to Softbank's USD 3.35 billion investment in Cruise Automation. This self-driving car company owned by General Motors develops autopilot systems for existing cars. In 2017, Ford invested USD 1 billion in AV company Argo AI.

## Broader trends in development and diffusion of AI

Efforts to develop empirical measures of AI are underway, but are challenged by definitional issues, among other concerns. Clear definitions are critical to compile accurate and comparable measures. Joint experimental work by the OECD and the Max Planck Institute for Innovation and Competition (MPI) proposes a three-pronged approach aimed at measuring i) AI developments in science, as captured by scientific publications; ii) technological developments in AI, as proxied by patents; and iii) AI software developments, in particular open-source software. The approach entails using expert advice to identify documents (publications, patents and software) that are unambiguously AI-related. These documents are then used as a benchmark to assess the degree of AI-relatedness of other documents (Baruffaldi et al., forthcoming[8]).

Scientific publications have long been used to proxy the outcome of research efforts and of advancements in science. The OECD uses bibliometric data from Scopus, a large abstracts and citations database of peer-reviewed literature and conference proceedings. Conference proceedings are particularly important in the case of emerging fields such as AI. They help provide a timely picture of new developments discussed at peer-reviewed conferences prior to being published. By establishing a list of AI-related keywords and validating them with AI experts, the approach aims to identify AI-related documents in any scientific domain.

The patent-based approach developed by the OECD and MPI Patent identifies and maps AI-related inventions and other technological developments that embed AI-related components in any technological domain. It uses a number of methods to identify AI inventions, including keyword search in patents' abstracts or claims; analysis of the patent portfolio of AI start-ups; and analysis of patents that cite AI-related scientific documents. This approach has been refined through work under the aegis of the OECD-led Intellectual Property (IP) Statistics Task Force.[1]

Data from GitHub – the largest open-source software hosting platform – are used to help identify AI developments. AI codes are divided into different topics with topic modelling analysis to show key AI fields. General fields comprise ML (including deep learning), statistics, mathematics and computational methods. Specific fields and applications include text mining, image recognition or biology.

# References

Agrawal, A., J. Gans and A. Goldfarb (2018), *Prediction Machines: The Simple Economics of Artificial Intelligence*, Harvard Business School Press, Brighton, MA.  [1]

Baruffaldi, S. et al. (forthcoming), "Identifying and measuring developments in artificial intelligence", *OECD Science, Technology and Industry Working Papers*, OECD Publishing, Paris.  [8]

Breschi, S., J. Lassébie and C. Menon (2018), "A portrait of innovative start-ups across countries", *OECD Science, Technology and Industry Working Papers*, No. 2018/2, OECD Publishing, Paris, http://dx.doi.org/10.1787/f9ff02f4-en.  [7]

Bresnahan, T. and M. Trajtenberg (1992), "General purpose technologies: 'Engines of growth?'", *NBER Working Paper*, No. 4148, http://dx.doi.org/10.3386/w4148.  [2]

Brynjolfsson, E., D. Rock and C. Syverson (2017), "Artificial intelligence and the modern productivity paradox: A clash of expectations and statistics", *NBER Working Paper*, No. 24001, http://dx.doi.org/10.3386/w24001.  [3]

CBI (2018), "The race for AI: Google, Intel, Apple in a rush to grab artificial intelligence startups", *CBI Insights,* 27 February, https://www.cbinsights.com/research/top-acquirers-ai-startups-ma-timeline/.  [6]

Dilda, V. (2017), *AI: Perspectives and Opportunities*, presentation at "AI: Intelligent Machines, Smart Policies" conference, Paris, 26-27 October, http://www.oecd.org/going-digital/ai-intelligent-machines-smart-policies/conference-agenda/ai-intelligent-machines-smart-policies-dilda.pdf.  [5]

MGI (2017), "Artificial Intelligence: The Next Digital Frontier?", Discussion Paper, McKinsey Global Institute, June, https://www.mckinsey.com/~/media/McKinsey/Industries/Advanced%20Electronics/Our%20Insights/How%20artificial%20intelligence%20can%20deliver%20real%20value%20to%20companies/MGI-Artificial-Intelligence-Discussion-paper.ashx.  [4]

## Note

[1] This took place with the advice of experts and patent examiners from the Australian IP Office, the Canadian Intellectual Property Office, the European Patent Office, the Israel Patent Office, the Italian Patent and Trademark Office, the National Institute for Industrial Property of Chile, the United Kingdom Intellectual Property Office and the United States Patent and Trademark Office.

# 3. AI applications

*This chapter illustrates opportunities in several sectors where artificial intelligence (AI) technologies are seeing rapid uptake, including transport, agriculture, finance, marketing and advertising, science, healthcare, criminal justice, security the public sector, as well as in augmented and virtual reality applications. In these sectors, AI systems can detect patterns in enormous volumes of data and model complex, interdependent systems to generate outcomes that improve the efficiency of decision making, save costs and enable better resource allocation. The section on AI in transportation was developed by the Massachusetts Institute of Technology's Internet Policy Research Institute. Several sections build on work being undertaken across the OECD, including the Committee on Digital Economy Policy and its Working Party on Privacy and Security, the Committee for Scientific and Technological Policy, the e-leaders initiative of the Public Governance Committee, as well as the Committee on Consumer Policy and its Working Party on Consumer Product Safety.*

## AI in transportation with autonomous vehicles

Artificial intelligence (AI) systems are emerging across the economy. However, one of the most transformational shifts has been with transportation and the transition to self-driving, or autonomous vehicles (AVs).

### Economic and social impact of AVs

Transportation is one of the largest sectors in economies across the OECD. In 2016, it accounted for 5.6% of gross domestic product across the OECD (OECD, 2018[1]).[1] The potential economic impact of introducing AVs into the economy could be significant due to savings from fewer crashes, less congestion and other benefits. It is estimated that a 10% adoption rate of AVs in the United States would save 1 100 lives and save USD 38 billion per year. A 90% adoption rate could save 21 700 lives and reduce annual costs by USD 447 billion (Fagnant and Kockelman, 2015[2]).

More recent research has found significant cost differences per kilometre for different transportation modes with and without vehicle automation in Switzerland (Bösch et al., 2018[3]). Their findings suggest that taxis will enjoy the largest cost savings. Individuals with private cars will receive smaller cost savings (Figure 3.1). Not surprisingly, the savings for taxis are largely due to elimination of driver wages.

**Figure 3.1. Cost comparison of different modes with and without AV technology**

In CHF per passenger kilometre

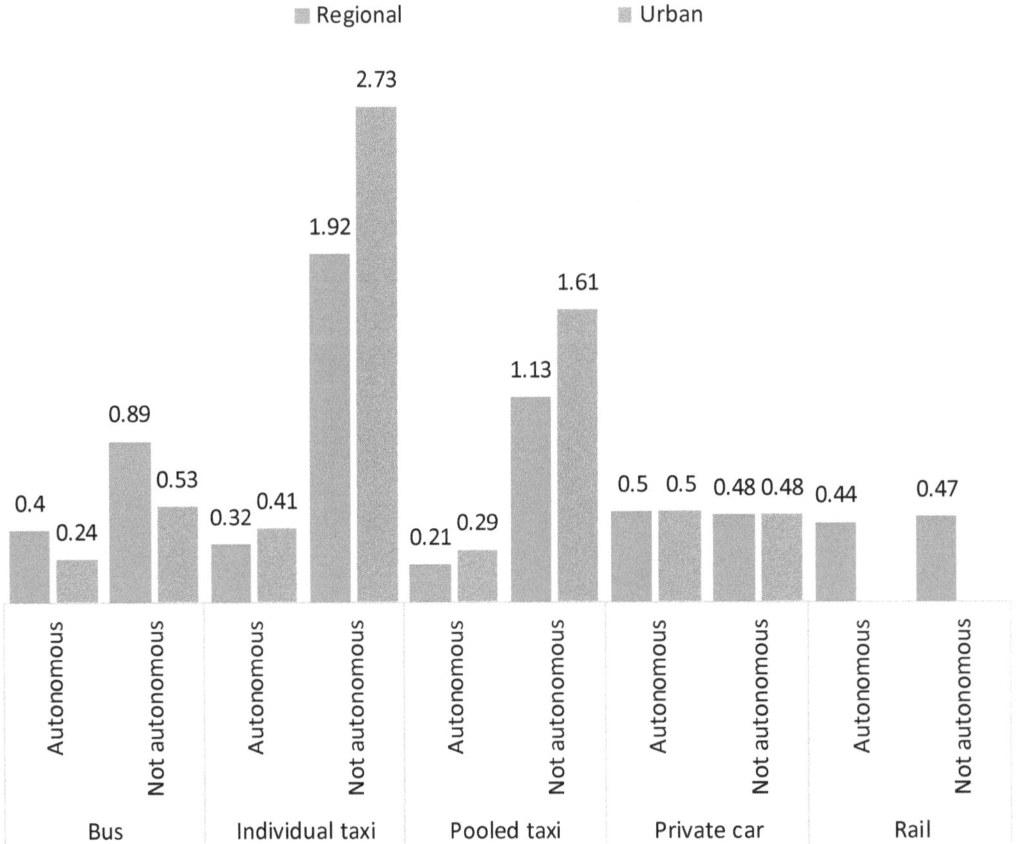

*Source*: Adapted from Bösch et al. (2018[3]), "Cost-based analysis of autonomous mobility services", https://doi.org/10.1016/j.tranpol.2017.09.005.

*Market evolution*

The state of transportation is in flux due to three significant and recent market shifts: the development of AV systems, the adoption of ride-sharing services and the shift to electric power vehicles. Traditional automobile manufacturers struggle to define their strategies in the face of two trends. First, ride-sharing services are increasing viable transportation options for users, particularly younger generations. Second, there are questions about the long-term viability of traditional car ownership. High-end manufacturers are already experimenting with new business models such as subscription services. Examples include "Access by BMW", "Mercedes Collection" and "Porsche Passport" where users can pay a flat monthly fee and exchange cars when they like.

Technology companies, from large multinationals to small start-ups, are moving into AV systems, ride-sharing services or electric vehicles – or some combination of the three. Morgan Stanley recently estimated Alphabet's Waymo division to be worth up to USD 175 billion on its own, based largely on its potential for autonomous trucking and delivery services (Ohnsman, 2018[4]). Zoox, a recent start-up focused on AI systems for driving in dense urban environments, has raised USD 790 million. This gives it a valuation of USD 3.2 billion[2] before producing any revenues (see also Section "Private equity investments in AI start-ups" in Chapter 2). These actions by technology companies complement the investment of traditional automakers and parts suppliers in AI-related technologies for vehicles.

**Figure 3.2. Patent filings related to AVs by company, 2011-16**

Companies with more than 50 patent fillings related to AVs

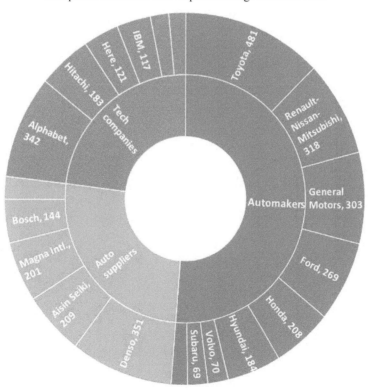

*Source*: Adapted from Lippert et al. (2018[5]), "Toyota's vision of autonomous cars is not exactly driverless", https://www.bloomberg.com/news/features/2018-09-19/toyota-s-vision-of-autonomous-cars-is-not-exactly-driverless.

Given the complexity of AV systems, companies tend to focus on their specific areas of expertise and then partner with firms specialising in others. Waymo is one of the leading firms in AV given its specialisation in massive data sets and ML. However, it does not build its own cars, choosing instead to rely on partners such as General Motors (GM) and Jaguar (Higgins and Dawson, 2018[6]).

Large auto manufacturers have also partnered with smaller start-ups to gain access to cutting-edge technology. For example, in October 2018 Honda announced a USD 2.75 billion investment in GM's Cruise self-driving venture (Carey and Lienert, 2018[7]). Ride-sharing firms such as Uber have also invested significantly in AVs and set up partnerships with leading technical universities (CMU, 2015[8]). This, however, has introduced questions of liability in the case of accidents, particularly when multiple stakeholders are in charge of multiple parts.

The diversity of market players investing in AV capabilities can be seen in the number of patent filings related to AVs by different groups of firms (Figure 3.2). Large automakers have considerable investments in intellectual property (IP); they are closely followed by auto suppliers and technology companies.

### Technology evolution

At the basic level, AVs have new systems of sensors and processing capacity that generate new complexities in the extract, transform and load process of their data systems. Innovation is flourishing amid high levels of investment in all key areas for AV. Less expensive light detection and ranging systems, for example, can map out the environment. In addition, new computer vision technologies can track the eyes and focus of drivers and determine when they are distracted. Now, after pulling in data and processing it, AI is adding another step: split-second operational decisions.

The core standard for measuring the progress of AV development is a six-stage standard developed by the Society of Automotive Engineers (SAE) (ORAD, 2016[9]). The levels can be summarised as follows:

**Level 0 (no driving automation)**: A human driver controls everything. There is no automated steering, acceleration, braking, etc.

**Level 1 (driver assistance)**: There is a basic level of automation, but the driver remains in control of most functions. The SAE says lateral (steering) or longitudinal control (e.g. acceleration) can be done autonomously, but not simultaneously, at this level.

**Level 2 (partial driving automation)**: Both lateral and longitudinal motion is controlled autonomously, for example with adaptive cruise control and functionality that keeps the car in its lane.

**Level 3 (conditional driving automation)**: A car can drive on its own, but needs to be able to tell the human driver when to take over. The driver is considered the fallback for the system and must stay alert and ready.

**Level 4 (high driving automation)**: The car can drive itself and does not rely on a human to take over in case of a problem. However, the system is not yet capable of autonomous driving in all circumstances (depending on situation, geographic area, etc.).

**Level 5 (full driving automation)**: The car can drive itself without any expectation of human intervention, and can be used in all driving situations. There is significant debate among stakeholders about how far the process has come towards fully autonomous driving. Stakeholders also disagree about the right approach to introduce autonomous functionality into vehicles.

Two key discussions focus on the role of the driver and the availability of the technology:

a) **The role of the driver**

   **Eliminating the need for a human driver**: Some firms developing AVs such as Waymo and Tesla believe it will soon be possible to eliminate the need for a human driver (owner or safety monitor). Tesla sells cars with level 3 autonomy. Waymo had plans to launch a fully autonomous taxi service with no driver in Arizona by the end of 2018 (Lee, 2018[10]).

   **Supporting the driver**: Other system developers believe the best use of AV systems for the near term will be to avoid accidents rather than to replace drivers. Toyota, the world's most valuable automaker by market capitalisation, is emphasising development of a vehicle that is incapable of causing a crash (Lippert et al., 2018[5]).

b) **The scope of availability**: There are two emerging approaches for initiating the deployment of automation in vehicles as described by Walker-Smith (2013[11]) and ITF (2018[12]).

   **Everything somewhere**: In this approach, very-high-level functionality is possible only in certain geographic areas or on certain roads that have been mapped in detail. Cadillac's Super Cruise, for example, is only available in certain places (e.g. it will only work on divided highways that have been mapped).

   **Something everywhere**: In this approach, functionality is only introduced to an AV system when it can be deployed on any road and in any situation. The result is a limited set of functions that should work in all locations. This appears to be the preferred approach of many automobile manufacturers.

Among more optimistic firms, the years 2020 and 2021 seem to be key targets for delivering AVs with level 4 functionality. Tesla and Zoox, for example, have set 2020 as a target, while Audi/Volkswagen, Baidu and Ford have targeted 2021. Renault Nissan has targeted 2022. Other manufacturers are also investing heavily in the technology. However, they are focused on preventing accidents from human drivers or believe the technology is not sufficiently developed for level 4 driving in the near team. These include BMW, Toyota, Volvo and Hyundai (Welsch and Behrmann, 2018[13]).

## Policy issues

The rollout of AVs raises a number of important legal and regulatory issues (Inners and Kun, 2017[14]). They concern specifically security and privacy (Bose et al., 2016[15]), but also touch more broadly on economy and society (Surakitbanharn et al., 2018[16]). Some more important areas of policy concern for OECD countries can be grouped as follows:

## Safety and regulation

In addition to ensuring safety (Subsection "Robustness, security and safety" in Chapter 4), policy issues include liability, equipment regulations for controls and signals, driver regulation and the consideration of traffic laws and operating rules (Inners and Kun, 2017[14]).

## Data

As with any AI system, access to data to train and adjust systems will be critical for the success of AVs. AV manufacturers have gathered immense data over the course of their trials. Fridman (8 October 2018[17]) estimates that Tesla has data for over 2.4 billion kilometres driven by its Autopilot. These real-time driving data that AV developers collect is proprietary and not shared across firms. However, initiatives such as the one by the Massachusetts Institute of Technology (MIT) (Fridman et al., 2018[18]) are building accessible data sets to understand driver behaviour. Their accessibility makes them particularly important for researchers and AV developers looking to improve systems. Policy discussions could include access to data collected by various systems and the government's role in funding open data collections.

### Security and privacy

AV systems require large amounts of data about the system, driver behaviour and their environment to function reliably and safely. These systems will also connect to various networks to relay information. The data collected, accessed and used by AV systems will need to be sufficiently secured against unwanted access. Such data can also include sensitive information such as location and user behaviour that will need to be managed and protected (Bose et al., 2016[15]). The International Transport Forum calls for comprehensive cybersecurity frameworks for automated driving (ITF, 2018[12]). New cryptographic protocols and systems also offer the promise of protecting privacy and securing the data. Yet these systems may slow down processing time for mission-critical and safety-critical tasks. In addition, they are in their early stages and not yet available at scales and speeds required by real-time AV deployments.

### Workforce disruption

The shift to AVs could have a significant effect on freight, taxi, delivery and other service jobs. In the United States, for example, an estimated 2.86% of workers have driving occupations (Surakitbharn et al., 2018[16]). Bösch et al. (2018[3]) highlight potentially significant cost savings in these industries from a shift to autonomous systems. Therefore, a rapid transition to AVs in the industry from a profit-maximising perspective might be expected when the technology is sufficiently advanced. Non-technical barriers such as regulation, however, would need to be overcome. This technological shift will displace workers, highlighting the need for policy work focused on skills and jobs in the context of a transitioning work environment (OECD, 2014[19]).

### Infrastructure

The introduction of AVs may require changes to infrastructure in keeping with the move to a mixed driving environment with a combination of human drivers and AVs. AVs may have the necessary equipment to communicate with each other in the future. However, legacy automobiles with human drivers would remain a significant source of uncertainty. AVs would need to adjust their behaviour in response to human-controlled vehicles. The possibility of dedicated AV lanes or infrastructure that could separate human drivers from AVs in the future is being discussed (Surakitbharn et al., 2018[16]). Infrastructure policy will need to integrate AV awareness into the planning process as the technology advances and AVs roll out.

## AI in agriculture

Improving accuracy of cognitive computing technologies such as image recognition is changing agriculture. Traditionally, agriculture has relied on the eyes and hands of experienced farmers to identify the right crops to pick. "Harvesting" robots equipped with AI technologies and data from cameras and sensors can now make this decision in real time. This type of robot can increasingly perform tasks that previously required human labour and knowledge.

Technology start-ups are creating innovative solutions leveraging AI in agriculture (FAO, 2017[20]). They can be categorised as follows (Table 3.1):

**Agricultural robots** handle essential agricultural tasks such as harvesting crops. Compared to human workers, these robots are increasingly fast and productive.

**Crop and soil monitoring** leverages computer vision and deep-learning algorithms to monitor crop and soil health. Monitoring has improved due to greater availability of satellite data (Figure 3.3).

**Predictive analytics** use ML models to track and predict the impact of environmental factors on crop yield.

### Figure 3.3. Examples of satellite data use for better monitoring

A. An ML tool analyses time series of satellite data for crop classification in Germany

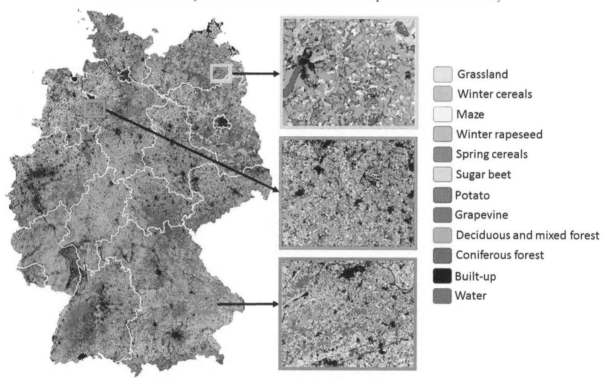

B. Ocean and coastal monitoring in Australia: seafloor (left) and turbidity (right)

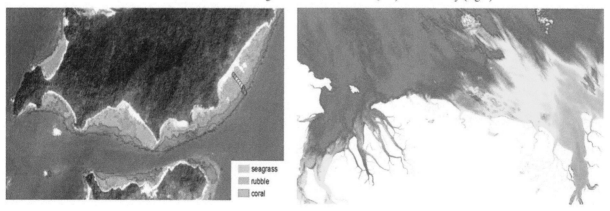

*Sources*: Roeland (2017[21]), *EC Perspectives on the Earth Observation*, www.oecd.org/going-digital/ai-intelligent-machines-smart-policies/conference-agenda/ai-intelligent-machines-smart-policies-roeland.pdf; Cooke (2017[22]), *Digital Earth Australia*, www.oecd.org/going-digital/ai-intelligent-machines-smart-policies/conference-agenda/ai-intelligent-machines-smart-policies-cooke.pdf.

**Table 3.1. A selection of AI start-ups in agriculture**

| Category | Company | Description |
|---|---|---|
| Agricultural robots | Abundant Robotics | Developed an apple-vacuum robot that uses computer vision to detect and pick apples with the same accuracy and care as a human. The company claims that the work of one robot is equivalent to that of ten people. |
| | Blue River Technology | Developed a robot known as See & Spray to monitor plants and soils and spray herbicide on weeds in lettuce and cotton fields. Precision spraying can help prevent resistance to herbicide and decrease the volume of chemicals used by 80%. John Deere acquired this company in September 2017 at USD 305 million. |
| | Harveset CROO Robotics | Developed a robot to help pick and pack strawberries. It can harvest 3.2 hectares a day and replace 30 human workers, helping to address labour shortage in key farming regions and prevent associated revenue losses. |
| Crop and soil monitoring | PEAT | Developed a deep-learning application to identify potential soil defects and nutrient deficiencies. It diagnoses plant health based on images taken by farmers. |
| | Resson | Developed image recognition algorithms that can accurately detect and classify plant pests and diseases. Resson has partnered with McCain Foods to help minimise losses in the potato production supply chain. |
| | SkySquirrel Technologies | Developed a system to analyse vineyard health based on images. Users upload images by drones to the company's cloud system that diagnoses grapevine leaves' condition. The company claims its technology can scan 20 hectares in 24 minutes and provide data analysis with 95% accuracy. |
| Predictive analytics | aWhere | Developed ML algorithms based on satellite data to predict weather conditions and provide customised advice to farmers, crop consultants and researchers. It also provides users with access to over a billion points of agronomic data daily. |
| | FarmShots | Developed a system to analyse agricultural data derived from satellite and drone images. The system can detect diseases, pests and poor plant nutrition on farms and inform users precisely where their fields need fertiliser, reducing the amount used by nearly 40%. |

*Source*: Companies' descriptions from their respective websites.

## *Challenges to AI adoption in agriculture*

The Food and Agriculture Organization of the United Nations (FAO) predicts the global population will grow by close to 30% between now and 2050 – from 7 billion to 9 billion people. However, only an additional 4% of land will be cultivated (FAO, 2009[23]). The OECD has investigated new opportunities and challenges of digital transformation in agriculture and the food sector (Jouanjean, 2019[24]). Among digital technologies, AI applications hold particular promise to increase agriculture productivity. However, the following challenges remain for wide adoption (Rakestraw, 2017[25]):

- **Lack of infrastructure**: Network connections remain poor in many rural areas. Also, data warehousing systems would be required to build robust applications.

- **Production of quality data**: AI applications in agriculture require high-quality data for recognition of crops or leaves. Collecting these data can be expensive because they can be captured only during the annual growing season.

- **Different mindset between tech start-ups and farmers**: Technology start-ups usually develop and launch products and services quickly, but farmers tend to adopt new processes and technologies more incrementally. Even big agricultural companies conduct lengthy field trials to ensure consistent performance and clear benefit of technology adoption.

- **Cost, notably for transactions**: High-tech farms (e.g. agricultural robots) require large investments in sensors and automation tools. For example, French agriculture is designing policies to encourage investment in specific AI agricultural applications. This could facilitate adoption of new technologies even for small-scale farmers (OECD, 2017[26]).

## *Potential ways to encourage adoption of AI in agriculture*

Solutions are being developed to address the various challenges to AI in agriculture. As in other areas of applications, open-source software is being developed and could help address cost issues. For example, Connectra has developed a motion-sensing device that attaches to a cow's neck and monitors its health based on Google's TensorFlow open-source software suite (Webb, 2017[27]). Transfer learning (see Subsection "Access and use of data" in Chapter 4) is helping address data issues by training algorithms with much smaller data sets. For example, researchers developed a system to detect diseases in the cassava plant that leverages learning from a different type of plant. With input of only 2 756 images of cassava leaves from plants in Tanzania, the ML researchers correctly identified brown leaf spot disease on cassava plants with 98% accuracy (Simon, 2017[28]).

## AI in financial services

In the financial sector, large companies such as JPMorgan, Citibank, State Farm and Liberty Mutual are rapidly deploying AI. The same is true for start-ups such as Zest Finance, Insurify, WeCash, CreditVidya and Aire. Financial service companies are combining different ML practices. For example, French start-up QuantCube Technology analyses several billion data points collected from over 40 countries. It uses language processing, deep learning, graph theory and more to develop AI solutions for decision making in financial corporates.

Deploying AI in the financial sector has many significant benefits. These include improving customer experience, identifying rapidly smart investment opportunities and possibly granting customers more credit with better conditions. However, it raises policy questions related to ensuring accuracy and preventing discrimination, as well as the broader impact of automation on jobs.

This section provides an overview of AI applications in the financial sector. It covers credit scoring, financial technology (FinTech), algorithmic trading, cost reduction in financial services, customer experience and compliance.

## *Credit scoring*

The financial services industry has long used statistical approaches for different ends, including calculating down-payment amounts and estimating risk of default. Credit scoring is a statistical analysis performed by financial institutions to assess a person's credit-worthiness. In other words, it assesses the possibility a borrower may default on her/his debt obligations. In traditional credit-scoring models, analysts make hypotheses regarding the attributes affecting a credit score and create customer segments.

More recently, neural network techniques have enabled the analysis of vast quantities of data collected from credit reports. They can conduct fine-grained analysis of the most relevant factors and of their relationships. In AI systems, algorithms based on large datasets automatically determine the leveraging of neural networks, customer segments and their weights. Credit bureaus in the United States report that deep-learning techniques that analyse data in new ways can improve accuracy of predictions by up to 15% (Press, 2017[29]).

As in other sectors, the difficulty to explain results from credit-scoring algorithms based on ML is an issue. Legal standards in several countries require high levels of transparency in the financial services sector. For example, in the United States, the Fair Credit Reporting Act (1970) and the Equal Credit Opportunity Act (1974) imply that the process and the output of any algorithm have to be explainable. Companies seem to be acting. For example,

Equifax, a credit reporting agency, and SAS, a data analysis company, have created an interpretable credit-scoring tool based on deep learning.

### *Financial technology lending*

FinTech businesses have grown rapidly in recent years. FinTech lending platforms allow consumers to shop for, apply and obtain loans online within seconds. They provide lenders with traditional credit report data (including payment history, amounts owed, length of history, number of accounts and more). In addition, FinTech lenders leverage a variety of alternative data sources. These include insurance claims, social media activities, online shopping information from marketplaces such as Amazon, shipping data from postal services, browsing patterns, and type of telephone or browser used (Jagtiani and Lemieux, 2019[30]). Research shows that alternative data processed by FinTech companies using AI can facilitate access to credit for those with no traditional credit history. They can also lower the costs associated with lending both for consumers and lenders (FSB, 2017[31]).

Research has compared the performance of algorithms to predict the probability of default based on the traditional FICO[3] score used in the United States and alternative data (Berg et al., 2018[32]). The FICO score alone had an accuracy rate of 68.3%, while an algorithm based on alternative data had an accuracy rate of 69.6%. Using both types of data together, the accuracy rate rose to 73.6%. These results suggest that alternative data complements, rather than substitutes for, credit bureau information. Thus, a lender using information from both traditional (FICO) sources, as well as alternative data, can make better lending decisions.

In the People's Republic of China (hereafter "China"), Ant Financial has emphasised how AI has driven its loan success (Zeng, 2018[33]). It uses algorithms to process the huge amount of transaction data generated by small businesses on its platform. This has allowed Ant to lend more than USD 13.4 billion to nearly 3 million small businesses. Ant's algorithms analyse transaction data automatically on all borrowers and on all their behavioural data in real time. It can process loans as small as several hundred Yuan renminbi (around USD 50) in a few minutes. Every action taken on Alibaba's platform – transaction, communication between seller and buyer, or connection with other services – affects a business's credit score. At the same time, the algorithms that calculate the scores themselves evolve over time, improving the quality of decision making with each iteration. The micro-lending operation has a default rate of about 1%, compared to the World Bank's 2016 estimate of an average of 4% worldwide.

Credit-scoring company Alipay uses consumer data points to determine credit scores (O'Dwyer, 2018[34]). These include purchase history, type of phone used, games played and friends on social media. In addition to traditional credit scoring to grant loans, the Chinese social credit score can influence decisions like the deposit level on an apartment rental or online dating matches. A person playing video games for hours every day might, for example, obtain a lower social credit score than a person purchasing diapers who is assumed to be a responsible parent (Rollet, 2018[35]). A broader Chinese social credit system to score the "trustworthiness" of individuals, businesses and government officials is planned to be in place by 2020.

Alternative data have the possibility of expanding access to credit. However, some caution that use of alternative data may raise concerns about disparate impact, privacy, security and "explainability" (Gordon and Stewart, 2017[36]). As a result, the Consumer Financial Protection Bureau in the United States investigated the use of alternative data in credit scoring (CFPB, 2017[37]).

## *Deploying AI for cost reduction in financial services*

The use of AI benefits both customers and financial institutions within the front office (e.g. client interaction), middle office (e.g. support for front office) and back office (e.g. settlements, human resources, compliance). The deployment of AI in the front, middle and back offices is expected to save financial entities an estimated 1 trillion dollars by 2030 in the United States, impacting 2.5 million financial services employees (Sokolin and Low, 2018[38]). Increasingly advanced AI tools are decreasing the need for human intervention.

In the front office, financial data and account actions are being integrated with AI-powered software agents. These agents can converse with clients within platforms such as Facebook Messenger or Slack that use advanced language processing. In addition to improving traditional customer service with AI, many financial companies are using AI to power "robot advisors". In this approach, algorithms provide automated financial advice and offerings (OECD, 2017[39]).

Another interesting development is the use of sentiment analysis on financial social media platforms. Companies such as Seeking Alpha and StockTwits focus on the stock market, enabling users to connect with each other and consult with professionals to grow their investment. The data produced on these platforms can be integrated in decision-making processes (Sohangir et al., 2018[40]). AI also helps enable online and mobile banking by authenticating users via fingerprint or facial recognition captured by smart phones. Alternatively, banks use voice recognition as a password to customer service rather than numerical passcodes (Sokolin and Low, 2018[38]).

In the middle office, AI can facilitate risk management and regulatory oversight processes. In addition, AI is helping portfolio managers to invest more efficiently and accurately. In back office product design, AI is broadening data sources to assess credit risk, take insurance underwriting risk and assess claims damage (e.g. assessing a broken windshield using machine vision).

## *Legal compliance*

The financial sector is well known for the high cost of complying with standards and regulatory reporting requirements. New regulation over the past decade in the United States and European Union has further heightened the cost of regulatory compliance for banks. In recent years, banks spent an estimated USD 70 billion annually on regulatory compliance and governance software. This spending reflects the cost of having bank attorneys, paralegals and other officers verify transaction compliance. Costs for these activities were expected to grow to nearly USD 120 billion in 2020 (Chintamaneni, 26 June 2017[41]). Deploying AI technologies, particularly language processing, is expected to decrease banks' compliance costs by approximately 30%. It will significantly decrease the time needed to verify each transaction. AI can help interpret regulatory documents and codify compliance rules. For example, the Coin program created by JPMorgan Chase reviews documents based on business rules and data validation. In seconds, the program can examine documents that would take a human being 360 000 hours of work to review (Song, 2017[42]).

## *Fraud detection*

Fraud detection is another major application of AI by financial companies. Banks have always monitored account activity patterns. Advances in ML, however, are starting to enable near real-time monitoring. This is allowing identification of anomalies immediately, which trigger a review. The ability of AI to continuously analyse new behaviour patterns and to automatically self-adjust is uniquely important for fraud detection because patterns evolve rapidly. In 2016, the bank Credit Suisse Group AG launched an AI joint venture with Silicon Valley surveillance and security firm Palantir Technologies. To help banks detect unauthorised trading, they developed a solution that aims to catch employees with unethical behaviours before they can harm the bank (Voegeli, 2016[43]). Fraud detection based on ML biometric security systems is also gaining traction in the telecommunications sector.

### Algorithmic trading

Algorithmic trading is the use of computer algorithms to decide on trades automatically, submit orders and manage those orders after submission. The popularity of algorithmic trading has grown dramatically over the past decade. It now accounts for the majority of trades put through exchanges globally. In 2017, JPMorgan estimated that just 10% of trading volume in stocks was "regular stock picking" (Cheng, 2017[44]). Increased computing capabilities enable "high frequency trading" whereby millions of orders are transmitted every day and many markets are scanned simultaneously. In addition, while most human brokers use the same type of predictors, the use of AI allows more factors to be considered.

## AI in marketing and advertising

AI is influencing marketing and advertising in many ways. At the core, AI is enabling the personalisation of online experiences. This helps display the content in which consumers are most likely to be interested. Developments in ML, coupled with the large quantities of data being generated, increasingly allow advertisers to target their campaigns. They can deliver personalised and dynamic ads to consumers at an unprecedented scale (Chow, 2017[45]). Personalised advertising offers significant benefits to enterprises and consumers. For enterprises, it could increase sales and the return on investment of marketing campaigns. For consumers, online services funded by advertising revenue are often provided free of charge to end users and can significantly decrease consumers' research costs.

The following non-exhaustive list outlines some developments in AI that could have a large impact on marketing and advertising practices around the world:

**Language processing**: One of the major subfields of AI that increases personalisation of ads and marketing messages is natural language processing (NLP). It enables the tailoring of marketing campaigns based on linguistic context such as social media posts, emails, customer service interactions and product reviews. Through NLP algorithms, machines learn words and identify patterns of words in common human language. They improve their accuracy as they go. In so doing, they can infer a customer's preferences and buying intent (Hinds, 2018[46]). NLP can improve the quality of online search results and create a better match between the customer's expectations and the ads presented, leading to greater advertising efficiency. For example, if customers searched online for a specific brand of shoes, an AI-based advertising algorithm could send targeted ads for this brand while they are doing unrelated tasks online. It can even send phone notifications when customers walk close to a shoe store offering discounts.

**Structured data analysis**: AI's marketing impact goes beyond the use of NLP models to analyse "unstructured data". Because of AI, today's online recommendation algorithms vastly outdo simple sets of guidelines or historical ratings from users. Instead, a wide range of data is used to provide customised recommendations. For instance, Netflix creates personalised suggested watching lists by considering what movies a person has watched or the ratings given to those movies. However, it also analyses which movies are watched multiple times, rewound and fast-forwarded (Plummer, 2017[47]).

**Determining the likelihood of success**: In online advertising, click-through rate (CTR) – the number of people who click on an ad divided by the number who have seen the ad – is an important metric for assessing ad performance. As a result, click prediction systems based on ML algorithms have been designed to maximise the impact of sponsored ads and online marketing campaigns. For the most part, Reinforced Learning algorithms are used to select the ad that incorporates the characteristics that would maximise CTR in the targeted population. Boosting CTR could significantly increase businesses' revenue: a 1% CTR improvement could yield huge gains in additional sales (Hong, 27 August 2017[48]).

**Personalised pricing**:[4] AI technologies are allowing companies to offer prices that continuously adjust to consumer behaviour and preferences. At the same time, companies can respond to the laws of supply and demand, profit requirements and external influences. ML algorithms can predict the top price a customer will pay for a product. These prices are uniquely tailored to the individual consumer at the point of engagement, such as online platforms (Waid, 2018[49]). On the one hand, AI can leverage dynamic pricing to the consumer's benefit. On the other, personalised pricing will likely be detrimental if it involves exploitative, distortionary or exclusionary pricing (Brodmerkel, 2017[50]).

**AI-powered augmented reality**: Augmented reality (AR) provides digital representations of products superimposed on the customer's view of the real world. AR combined with AI can give customers an idea of how the product would look once produced and placed in its projected physical context. AI-powered AR systems can learn from a customer's preferences. It can then adapt the computer-generated images of the products accordingly, improving customer experience and increasing the likelihood of buying (De Jesus, 2018[51]). AR could expand the online shopping market and thus boost online advertising revenue.

## AI in science

Global challenges today range from climate change to antibiotic bacterial resistance. Solutions to many of these challenges require increases in scientific knowledge. AI could increase the productivity of science, at a time when some scholars are claiming that new ideas may be becoming harder to find (Bloom et al., 2017[52]). AI also promises to improve research productivity even as pressure on public research budgets is increasing. Scientific insight depends on drawing understanding from vast amounts of scientific data generated by new scientific instrumentation. In this context, using AI in science is becoming indispensable. Furthermore, AI will be a necessary complement to human scientists because the volume of scientific papers is vast and growing rapidly, and scientists may have reached "peak reading".[5]

The use of AI in science may also enable novel forms of discovery and enhance the reproducibility of scientific research. AI's applications in science and industry have become numerous and increasingly significant. For instance, AI has predicted the behaviour of chaotic systems, tackled complex computational problems in genetics, improved the quality of astronomical imaging and helped discover the rules of chemical synthesis. In addition, AI is being deployed in functions that range from analysis of large datasets, hypothesis generation, and comprehension and analysis of scientific literature to facilitation of data gathering, experimental design and experimentation itself.

### Recent drivers of AI in science

Forms of AI have been applied to scientific discovery for some time, even if this has been sporadic. For example, the AI program DENDRAL was used in the 1960s to help identify chemical structures. In the 1970s, an AI known as Automated Mathematician assisted mathematical research. Since those early approaches, computer hardware and software have vastly improved, and data availability has increased significantly. Several additional factors are also enabling AI in science: AI is well-funded, at least in the commercial sector; scientific data are increasingly abundant; high-performance computing is improving; and scientists now have access to open-source AI code.

### The diversity of AI applications in science

AI is in use across many fields of research. It is a frequently used technique in particle physics, for example, which depends on finding complex spatial patterns in vast streams of data yielded by particle detectors. With data gleaned from social media, AI is providing evidence on relationships between language use, psychology and health, and social and economic outcomes. AI is also tackling complex computational problems in genetics, improving the quality of imaging in astronomy and helping discover the rules of chemical synthesis, among other uses (OECD, 2018[53]). The range and frequency of such applications is likely to grow. As advances occur in automated ML process, scientists, businesses and other users can more readily employ this technology.

Progress has also occurred in AI-enabled hypothesis generation. For example, IBM has produced a prototype system, KnIT, which mines information contained in scientific literature. It represents this information explicitly in a queryable network, and reasons on these data to generate new and testable hypotheses. KnIT has text-mined published literature to identify new kinases – an enzyme that catalyses the transfer of phosphate groups from high-energy, phosphate-donating molecules to specific substrates. These kinases have introduced a phosphate group into a protein tumour suppressor (Spangler et al., 2014[54]).

AI is likewise assisting in the review, comprehension and analysis of scientific literature. NLP can now automatically extract both relationships and context from scientific papers. For example, the KnIT system involves automated hypothesis generation based on text mining of scientific literature. Iris.AI[6] is a start-up that offers a free tool to extract key concepts from research abstracts. It presents the concepts visually (such that the user can see cross-disciplinary relationships). It also gathers relevant papers from a library of over 66 million open access papers.

AI is assisting in large-scale data collection. In citizen science, for example, applications use AI to help users identify unknown animal and plant specimens (Matchar, 2017[55]).

### AI can also combine with robotic systems to execute closed-loop scientific research

The convergence of AI and robotics has many potential benefits for science. Laboratory-automation systems can physically exploit techniques from the AI field to pursue scientific experiments. At a laboratory at the University of Aberystwyth in Wales, for example, a robot named Adam uses AI techniques to perform cycles of scientific experimentation automatically. It has been described as the first machine to independently discover new scientific knowledge. Specifically, it discovered a compound, Triclosan, that works against wild-type and drug-resistant *Plasmodium falciparum* and *Plasmodium vivax* (King et al., 2004[56]). Fully automating science has several potential advantages (OECD, 2018[57]):

- **Faster scientific discovery**: Automated systems can generate and test thousands of hypotheses in parallel. Due to their cognitive limits, human beings can only consider a few hypotheses at a time (King et al., 2004[56]).

- **Cheaper experimentation**: AI systems can select experiments that cost less to perform (Williams et al., 2015[58]). The power of AI offers efficient exploration and exploitation of unknown experimental landscapes. It leads the development of novel drugs (Segler, Preuss and Waller, 2018[59]), materials (Butler et al., 2018[60]) and devices (Kim et al., 2017[61]).

- **Easier training**: Including initial education, a human scientist requires over 20 years and huge resources to be fully trained. Humans can only absorb knowledge slowly through teaching and experience. Robots, by contrast, can directly absorb knowledge from each other.

- **Improved knowledge and data sharing and scientific reproducibility**: One of the most important issues in biology – and other scientific fields – is reproducibility. Robots have the superhuman ability to record experimental actions and results. These results, along with the associated metadata and employed procedures, are automatically and completely recorded and in accordance with accepted standards at no additional cost. By contrast, recording data, metadata and procedures adds up to 15% to the total costs of experimentation by humans.

Laboratory automation is essential to most areas of science and technology. However, it is expensive and difficult to use due to a low number of units sold and market immaturity. Consequently, laboratory automation is used most economically in large central sites. Indeed, companies and universities are increasingly concentrating their laboratory automation. The most advanced example of this trend is cloud automation. In this practice, a large amount of equipment is gathered in a single site. Biologists, for example, send their samples and use an application to help design their experiments.

## *Policy considerations*

The increasing use of AI systems in science could also affect sociological, institutional and other aspects of science. These include the transmission of knowledge, systems of credit for scientific discoveries, the peer-review system and systems of intellectual property rights. As AI contributes increasingly to the world of science, the importance of policies that affect access to data and high-performance computing will amplify. The growing prominence of AI in discovery is raising new, and as yet unanswered, questions. Should machines be included in academic citations? Will IP systems need adjustments in a world in which machines can invent? In addition, a fundamental policy issue concerns education and training (OECD, 2018[57]).

## AI in health

### *Background*

AI applications in healthcare and pharmaceuticals can help detect health conditions early, deliver preventative services, optimise clinical decision making, and discover new treatments and medications. They can facilitate personalised healthcare and precision medicine, while powering self-monitoring tools, applications and trackers. AI in healthcare offers potential benefits for quality and cost of care. Nevertheless, it also raises policy questions, in particular concerning access to (health) data (Section "AI in Health") and privacy (Subsection "Personal data protection" in Chapter 4). This section focuses on AI's specific implications for healthcare.

In some ways, the health sector is an ideal platform for AI systems and a perfect illustration of its potential impacts. A knowledge-intensive industry, it depends on data and analytics to improve therapies and practices. There has been tremendous growth in the range of information collected, including clinical, genetic, behavioural and environmental data. Every day, healthcare professionals, biomedical researchers and patients produce vast amounts of data from an array of devices. These include electronic health records (EHRs), genome sequencing machines, high-resolution medical imaging, smartphone applications and ubiquitous sensing, as well as Internet of Things (IoT) devices that monitor patient health (OECD, 2015[62]).

### Beneficial impact of AI on healthcare

If put to use, AI data generated could be of great value to healthcare and research. Indeed, health sectors across countries are undergoing a profound transformation as they capitalise on opportunities provided by information and communication technologies. Key objectives shaping this transformation process include improved efficiency, productivity and quality of care (OECD, 2017[26]).

### Specific illustrations

**Improving patient care**: Secondary use of health data can improve the quality and effectiveness of patient care, in both clinical and homecare settings. For example, AI systems can alert administrators and front-line clinicians when measures related to quality and patient safety fall outside a normal range. They can also highlight factors that may be contributing to the deviations (Canadian Institute for Health Information, 2013[63]). A specific aspect of improving patient care concerns **precision medicine**. This is based on rapid processing of a variety of complex datasets such as a patient's health records, physiological reactions and genomic data. Another aspect concerns **mobile health**: mobile technologies provide helpful real-time feedback along the care continuum – from prevention to diagnosis, treatment and monitoring. Linked with other personal information such as location and preferences, AI-enhanced technologies can identify risky behaviours or encourage beneficial ones. Thus, they can produce tailored interventions to promote healthier behaviour (e.g. taking the stairs instead of the lift, drinking water or walking more) and achieve better health outcomes. These technologies, as well as sensor-based monitoring systems, offer continuous and direct monitoring and personalised intervention. As such, they can be particularly useful to improve the quality of elderly care and the care of people with disabilities (OECD, 2015[62]).

**Managing health systems**: Health data can inform decisions regarding programmes, policy and funding. In this way, they can help manage and improve the effectiveness and efficiency of the health system. For example, AI systems can reduce costs by identifying ineffective interventions, missed opportunities and duplicated services. Access to care can be increased and wait times reduced through four key ways. First, AI systems understand patient journeys across the continuum of care. Second, they ensure that patients receive the services most appropriate for their needs. Third, they accurately project future healthcare needs of the population. Fourth, they optimise allocation of resources across the system (Canadian Institute for Health Information, 2013[63]). With increasing monitoring of therapies and events related to pharmaceuticals and medical devices (OECD, 2015[62]), countries can use AI to advance identification of patterns, such as systemic failures and successes. More generally, data-driven innovation fosters a vision for a "learning health system". Such a system can continuously incorporate data from researchers, providers and patients. This allows it to improve comprehensive clinical algorithms, reflecting preferred care at a series of decision nodes for clinical decision support (OECD, 2015[62]).

**Understanding and managing population and public health**: In addition to timelier public health surveillance of influenza and other viral outbreaks, data can be used to identify unanticipated side effects and contraindications of new drugs (Canadian Institute for Health Information, 2013[63]). AI technologies may allow for early identification of outbreaks and surveillance of disease spreading. Social media, for example, can both detect and disseminate information on public health. AI uses NLP tools to process posts on social media to extract potential side effects (Comfort et al., 2018[64]; Patton, 2018[65]).

**Facilitating health research**: Health data can support clinical research and accelerate discovery of new therapies. Big data analytics offers new and more powerful opportunities to measure disease progression and health for improved diagnosis and care delivery, as well as translational and clinical research, e.g. for developing new drugs. In 2015, for example, the pharmaceutical company Atomwise collaborated with researchers at the University of Toronto and IBM to use AI technology in performing Ebola treatment research.[7] The use of AI is also increasingly tested in medical diagnosis, with a landmark approval by the United States Food and Drug Administration. The ruling allowed marketing of the first medical device to use AI to "detect greater than a mild level of the eye disease diabetic retinopathy in adults who have diabetes" (FDA, 2018[66]). Similarly, ML techniques can be used to train models to classify images of the eye, potentially embedding cataract detectors in smartphones and bringing them to remote areas (Lee, Baughman and Lee, 2017[67]; Patton, 2018[65]). In a recent study, a deep-learning algorithm was fed more than 100 000 images of malignant melanomas and benign moles. It eventually outperformed a group of 58 international dermatologists in the detection of skin cancer (Mar and Soyer, 2018[68]).

### *Enabling AI in healthcare – success and risk factors*

Sufficient infrastructure and risk mitigation should be in place to take full advantage of AI capabilities in the health sector.

Countries are increasingly establishing EHR systems and adopting mobile health (m-health), allowing mobile services to support the practice of medicine and public health (OECD, 2017[69]). Robust evidence demonstrates how EHRs can help reduce medication errors and better co-ordinate care (OECD, 2017[26]). On the other hand, the same study showed that only a few countries have achieved high-level integration and capitalised on the possibility of extracting data from EHRs for research, statistics and other secondary uses. Healthcare systems still tend to capture data in silos and analyse them separately. Standards and interoperability are key challenges that must be addressed to realise the full potential of EHRs (OECD, 2017[26]).

Another critical factor for the use of AI in the health sector concerns minimising the **risks to data subjects' privacy**. The risks in increased collection and processing of personal data are described in detail in Subsection "Personal data protection" of Chapter 4. This subsection addresses the high sensitivity of health-related information. Bias in the operation of an algorithm recommending specific treatment could create real health risks to certain groups. Other privacy risks are particular to the health sector. For example, questions from the use of data extracted from implantable healthcare devices, such as pacemakers, could be evidenced in court.[8] Additionally, as these devices become more sophisticated, they raise increasing safety risks, such as a malicious takeover that would administer a harmful operation. Another example is the use of biological samples (e.g. tissues) for ML, which raises complex questions of consent and ownership (OECD, 2015[62]; Ornstein and Thomas, 2018[70]).[9]

As a result of these concerns, many OECD countries report legislative barriers to the use of personal health data. These barriers include disabling data linkages and hindering the development of databases from EHRs. The 2016 Recommendation of the Council on

Health Data Governance is an important step towards a more coherent approach in health data management and use (OECD, 2016[71]). It aims primarily to promote the establishment and implementation of a national health data governance framework. Such a framework would encourage the availability and use of personal health data to serve health-related public interests. At the same time, it would promote protection of privacy, personal health data and data security. Adopting a coherent approach to data management could help remove the trade-off between data use and security.

Involving all relevant stakeholders is an important means of garnering trust and public support in the use of AI and data collection for health purposes. Similarly, governments could develop appropriate trainings for future health data scientists, or pair data scientists with healthcare practitioners. In this way, they could provide better understanding of the opportunities and risks in this emerging field (OECD, 2015[62]). Involving clinicians in the design and development of AI healthcare systems could prove essential for getting patients and providers to trust AI-based healthcare products and services.

## AI in criminal justice

### *AI and predictive algorithms in the legal system*

AI holds the potential to improve access to justice and advance its effective and impartial adjudication. However, concerns exist about AI systems' potential challenges to citizen participation, transparency, dignity, privacy and liberty. This section will focus on AI advancement in the area of criminal justice, touching upon developments in other legal areas as well.

AI is increasingly used in different stages of the criminal procedure. These range from predicting where crimes may occur and the outcome of a criminal procedure to conducting risk assessments on defendants, as well as to contributing to more efficient management of the process. Although many AI applications are still experimental, a few advanced prediction products are already in use in justice administration and law enforcement. AI can improve the ability to make connections, detect patterns, and prevent and solve crimes (Wyllie, 2013[72]). The uptick in the use of such tools follows a larger trend of turning to fact-based methods as a more efficient, rational and cost-effective way to allocate scarce law enforcement resources (Horgan, 2008[73]).

Criminal justice is a sensitive point of interaction between governments and citizens, where asymmetry of power relations and information is particularly pronounced. Without sufficient safeguards, it might create disproportionately adverse results, reinforce systemic biases and possibly even create new ones (Barocas and Selbst, 2016[74]).

### *Predictive policing*

In predictive policing, law enforcement uses AI to identify patterns in order to make statistical predictions about potential criminal activity (Ferguson, 2014[75]). Predictive methods were used in policing even before the introduction of AI to this field. In one notable example, police analysed accumulated data to map cities into high- and low-risk neighbourhoods (Brayne, Rosenblat and Boyd, 2015[76]). However, AI can link multiple datasets and perform complex and more fine-grained analytics, thus providing more accurate predictions. For example, the combination of automatic license plate readers, ubiquitous cameras, inexpensive data storage and enhanced computing capabilities can provide police forces with significant information on many people. Using these data, police can identify patterns, including patterns of criminal behaviour (Joh, 2017[77]).

There are two major methods of predictive policing. **Location prediction** applies retrospective crime data to forecast when and where crimes are likely to occur. Locations could include liquor stores, bars and parks where certain crimes have occurred in the past. Law enforcement could attempt to prevent future crimes by deploying an officer to patrol these areas, on a specific day/time of the week. In **person-based prediction**, law enforcement departments use crime statistics to help predict which individuals or groups are most likely to be involved in crimes, either as victims or offenders.

AI-enhanced predictive policing initiatives are being trialled in cities around the world, including in Manchester, Durham, Bogota, London, Madrid, Copenhagen and Singapore. In the United Kingdom, the Greater Manchester Police developed a predictive crime mapping system in 2012. Since 2013, the Kent Police has been using a system called PredPol. These two systems estimate the likelihood of crime in particular locations during a window of time. They use an algorithm originally developed to predict earthquakes.

In Colombia, the Data-Pop Alliance uses crime and transportation data to predict criminal hotspots in Bogota. Police forces are then deployed to specific places and at specific times where risk of crime is higher.

Many police departments also rely on social media for a wide range of purposes. These include discovering criminal activity, obtaining probable cause for search warrants, collecting evidence for court hearings, pinpointing the locations of criminals, managing volatile situations, identifying witnesses, broadcasting information and soliciting tips from the public (Mateescu et al., 2015[78]).

The use of AI raises issues with respect to use of personal data (Subsection "Personal data protection" in Chapter 4) and to risks of bias (Subsection "Fairness and ethics" in Chapter 4). In particular, it raises concerns with regard to transparency and the ability to understand its operation. These issues are especially sensitive when it comes to criminal justice. One approach to improve algorithmic transparency, applied in the United Kingdom, is a framework called ALGO-CARE. This aims to ensure that police using algorithmic risk assessment tools consider key legal and practical elements (Burgess, 2018[79]). The initiative translates key public law and human rights principles, developed in high-level documents, into practical terms and guidance for police agencies.

## *Use of AI by the judiciary*

In several jurisdictions, the judiciary uses AI primarily to assess risk. Risk assessment informs an array of criminal justice outcomes such as the amount of bail or other conditions for release and the eligibility for parole (Kehl, Guo and Kessler, 2017[80]). The use of AI for risk assessment follows other forms of actuarial tools that judges have relied on for decades (Christin, Rosenblat and Boyd, 2015[81]). Researchers at the Berkman Klein Center at Harvard University are working on a database of all risk assessment tools used in the criminal justice systems in the United States to help inform decision making (Bavitz and Hessekiel, 2018[82]).

Risk assessment algorithms predict the risk level based on a small number of factors, typically divided into two groups. These are criminal history (e.g. previous arrests and convictions, and prior failures to appear in court) and sociodemographic characteristics (e.g. age, sex, employment and residence status). Predictive algorithms summarise the relevant information for making decisions more efficiently than the human brain. This is because they process more data at a faster rate and may also be less exposed to human prejudice (Christin, Rosenblat and Boyd, 2015[81]).

AI-based risk assessment tools developed by private companies raise unique transparency and explainability concerns. These arise because non-disclosure agreements often prevent access to proprietary code to protect IP or prevent access for malicious purposes (Joh, 2017[77]). Without access to the code, there are only limited ways to examine the validity and reliability of the tools.

The non-profit news organisation ProPublica reportedly tested the validity of a proprietary tool called COMPAS, which is used in some jurisdictions in the United States. It found that COMPAS predictions were accurate 60% of the time across all types of crime. However, the prediction accuracy rate for violent crime was only 20%. In addition, the study pointed out racial disparities. The algorithm falsely flagged black defendants as future criminals twice as often as it did with white defendants (Angwin et al., 2016[83]). The study attracted media attention and its results were questioned on the basis of statistical errors (Flores, Bechtel and Lowenkamp, 2016[84]). COMPAS is a "black box" algorithm, meaning that no one, including its operators, has access to the source code.

The use of COMPAS was challenged in court with opponents claiming its proprietary nature violates defendants' right to due process. The Supreme Court of Wisconsin approved the use of COMPAS in sentencing. However, it must remain an assistive tool and the judge must retain full discretion to determine additional factors and weigh them accordingly.[10] The US Supreme Court denied a petition to hear the case.[11]

In another study examining the impact of AI on criminal justice, Kleinberg et al. (2017[85]) built an ML algorithm. It aims to predict which defendants would commit an additional crime while awaiting trial or try to escape court (pre-trial failures). Input variables were known and the algorithm determined the relevant sub-categories and their respective weight. For the age variable, for example, the algorithm determined the most statistical significant division of age group brackets, such as 18-25 and 25-30 years old. The authors found this algorithm could considerably reduce incarceration rates, as well as racial disparities. Moreover, AI reduced human biases: the researchers concluded that any information beyond the necessary factors for prediction could distract judges and increase the risk of biased rulings.

Advanced AI-enhanced tools for risk assessment are also used in the United Kingdom. Durham Constabulary has developed the Harm Assessment Risk Tool to evaluate the risk of convicts reoffending. The tool is based on a person's past offending history, age, postcode and other background characteristics. Based on these indicators, algorithms classify the person as low, medium or high risk.

### *Using AI to predict the outcome of cases*

Using advanced language processing techniques and data analysis capabilities, several researchers have built algorithms to predict the outcome of cases with high accuracy rates. For example, researchers at University College London and the Universities of Sheffield and Pennsylvania, developed an ML algorithm that can predict the outcome of cases heard by the European Court of Human Rights with a 79% accuracy rate (Aletras et al., 2016[86]). Another study by researchers from the Illinois Institute of Technology in Chicago built an algorithm that can predict the outcome of cases brought before the US Supreme Court with a 79% accurate rate (Hutson, 2017[87]). The development of such algorithms could help the parties assess the likelihood of success in trial or on appeal (based on previous similar cases). It could also help lawyers identify which issues to highlight in order to increase their chances of winning.

## *Other uses of AI in legal procedures*

In civil cases, the use of AI is broader. Attorneys are using AI for drafting contracts, mining documents in discovery and due diligence (Marr, 2018[88]). The use of AI might expand to other similar areas of the criminal justice system such as plea bargain and examination. Because the design of the algorithms and their use could affect the outcome, policy implications related to AI need to be considered carefully.

## AI in security

AI promises to help address complex digital and physical security challenges. In 2018, global defence spending is forecasted to reach USD 1.67 trillion, a 3.3% year-on-year increase (IHS, 2017[89]). Security spending is not limited to the public sector, however. The private sector worldwide was expected to spend USD 96 billion to respond to security risks in 2018, an 8% increase from 2017 (Gartner, 2017[90]). Recent large-scale digital security attacks have increased society's awareness of digital security. They have demonstrated that data breaches can have far-reaching economic, social and national security consequences. Against this backdrop, public and private actors alike are adopting and employing AI technologies to adjust to the changing security landscape worldwide. This section describes two security-related areas that are experiencing particularly rapid uptake: digital security and surveillance.[12,13]

## *AI in digital security*

AI is already broadly used in digital security applications such as network security, anomaly detection, security operations automation and threat detection (OECD, 2017[26]). At the same time, malicious use of AI is expected to increase. Such malicious activities include identifying software vulnerabilities with the goal of exploiting them to breach the availability, integrity or confidentiality of systems, networks and data. This will affect the nature and overall level of digital security risk.

Two trends make AI systems increasingly relevant for security: the growing number of digital security attacks and the skills shortage in the digital security industry (ISACA, 2016[91]). As a result of these trends, ML tools and AI systems are becoming increasingly relevant to automate threat detection and response (MIT, 2018[92]). Malware constantly evolves. ML has become indispensable to combat attacks such as polymorphic viruses, denial of service and phishing.[14] Indeed, leading email service providers, such as Gmail and Outlook, have employed ML at varying levels of success for more than a decade to filter unwanted or pernicious email messages. Box 3.1 illustrates some uses of AI to protect enterprises against malicious threats.

Computer code is prone to human error. Nine out of ten digital security attacks are estimated to result from flaws in software code. This occurs despite the vast amount of development time – between 50% and 75% – spent on testing (FT, 2018[93]). Given the billions of lines of code being written every year and the re-use of third party proprietary libraries to do it, detecting and correcting errors in software code is a daunting task for the human eye. Countries such as the United States and China are funding research projects to make AI systems that can detect software security vulnerabilities. Companies such as Ubisoft – the video game maker – are starting to use AI to flag faulty code before it is implemented, effectively reducing testing time by 20% (FT, 2018[93]). In practice, software-checking AI technologies work like spellcheck tools that identify typos and syntax errors in word processing software. However, AI technologies learn and become more effective as they go (FT, 2018[93]).

---

**Box 3.1. Using AI to manage digital security risk in business environments**

Companies like Darktrace, Vectra and many others apply ML and AI to detect and react to digital security attacks in real time. Darktrace relies on Enterprise Immune System technology, which does not require previous experience of a threat to understand its potential danger. AI algorithms iteratively learn a network's unique "pattern of life" or "self" to spot emerging threats that would otherwise go unnoticed. In its methods, Darktrace is analogous to the human immune system, which learns about what is normal to the body and automatically identifies and neutralises situations outside such pattern of normality.

Vectra proactively hunts down attackers in cloud environments by using a non-stop, automated and always-learning "cognito platform". This provides full visibility into the attacker behaviours from cloud and data centre workloads to user and IoT devices. In this way, Vectra makes it increasingly hard for attackers to hide.

*Sources*: www.darktrace.com/; https://vectra.ai/.

---

## AI in surveillance

Digital infrastructure is developing in cities. This is especially true in the surveillance sector, where various tools that use AI are being installed to increase public security. Smart cameras, for example, can detect a fight. Gunshot locators automatically report recorded shots and provide the exact location. This section looks at how AI is revolutionising the world of public security and surveillance.

Video surveillance has become an increasingly common tool to enhance public security. In the United Kingdom, a recent study estimated that security footage provided useful evidence for 65% of the crimes committed on the British railway network between 2011 and 2015 for which footage was available (Ashby, 2017[94]). The massive volume of surveillance cameras – 245 million globally in 2014 – implies a growing amount of data being generated. This went from 413 Petabytes (PB) of information produced in just one day in 2013 to a daily estimate of 860 PB in 2017 (Jenkins, 2015[95]); (Civardi, 2017[96]). Humans have limited ability to process such high amounts of data. This gives way to the use of AI technologies designed to handle large volumes of data and automate mechanical processes of detection and control. Moreover, AI enables security systems to detect and react to crime in real time (Box 3.2).

---

**Box 3.2. Surveillance with "smart" cameras**

The French Commission for Atomic and Alternative Energies, in partnership with Thales, uses deep learning to automatically analyse and interpret videos for security applications. A Violent Event Detection module automatically detects violent interactions such as a fight or aggression captured by closed-circuit television cameras and alerts operators in real time. Another module helps locate the perpetrators on the camera network. These applications are being evaluated by French public transportation bodies RATP and SNCF in the Châtelet-Les Halles and Gare du Nord stations, two of Paris' busiest train and subway stations. The city of Toulouse, France, also uses smart cameras to signal unusual behaviour and spot abandoned luggage in public places. Similar projects are being trialled in Berlin, Rotterdam and Shanghai.

*Source*: Demonstrations and information provided to the OECD by CEA Tech and Thales in 2018. More information (in French) available at: www.gouvernement.fr/sites/default/files/contenu/piece-jointe/2015/11/projet_voie_videoprotection_ouverte_et_integree_appel_a_projets.pdf.

---

---

**Box 3.3. Face recognition as a tool for surveillance**

Face-recognition technologies are increasingly being used to provide effective surveillance by private or public actors (Figure 3.4). AI improves traditional face-recognition systems by allowing for faster and more accurate identification in cases where traditional systems would fail, such as poor lighting and obstructed targets. Companies such as FaceFirst combine face-recognition tools with AI to offer solutions to prevent theft, fraud and violence. Specific considerations are embedded into their design. This allows the design to meet the highest standards of privacy and security, such as anti-profiling to prevent discrimination, encryption of image data and strict timeframes for data purging. These surveillance tools have been applied across industries, ranging from retail (e.g. to stop shoplifting), banking (e.g. to prevent identity fraud) and law enforcement (e.g. for border security) to event management (e.g. to recognise banned fans) and casinos (e.g. to spot important people).

**Figure 3.4. Illustration of face-recognition software**

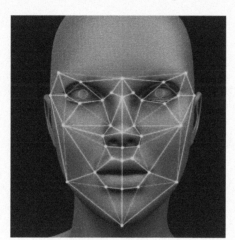

*Source*: www.facefirst.com.

---

In line with the dual-use nature of AI, surveillance tools incorporating AI could have illegitimate purposes that may go against the principles described in Chapter 4. Legitimate purposes include law enforcement to streamline criminal investigations, detect and stop crimes at their early stages and counter terrorism. Face-recognition technologies have proven relevant in this regard (Box 3.3). However, the impact of AI in surveillance goes beyond face-recognition systems. It also plays an increasingly important role in enhancing faceless recognition technologies. In these cases, alternative information about subjects (height, clothing, build, postures, etc.) is used for their identification. Additionally, AI has been effective when combined with image-sharpening technologies: large image datasets are used to train neural networks on the typical features of physical objects such as skin, hair or even bricks in a wall. The system then recognises such features in new images and adds extra details and textures to them using the knowledge previously acquired. This fills the gaps produced by poor image resolution, improving the effectiveness of surveillance systems (Medhi, Scholkopf and Hirsch, 2017[97]).

## AI in the public sector

The potential of AI for public administrations is manifold. The development of AI technologies is already having an impact on how the public sector works and designs policies to serve citizens and businesses. Applications touch on areas such as health, transportation and security services.[15]

Governments in OECD countries are experimenting with and implementing projects aimed at exploiting AI to better meet the needs of public-service users. They also want to enhance stewardship of their resources (e.g. increasingly saving the time civil servants spend on customer support and administrative tasks). AI tools could enhance the efficiency and quality of many public sector procedures. For example, they could offer citizens the opportunity to be engaged right up-front in the process of service design and to interact with the state in a more agile, effective and personalised way. If correctly designed and implemented, AI technologies could be integrated into the entire policy-making process, support public sector reforms and improve public sector productivity.

Some governments have deployed AI systems to strengthen social welfare programmes. For instance, AI could help attain optimal inventory levels at health and social service locations. They would do this through ML technologies that analyse transaction data and make increasingly accurate replenishment predictions. This, in turn, would facilitate forecasting and policy development. In another example, AI algorithms are helping the UK government detect fraud in social benefits claims (Marr, 2018[98]).

## AI applications using augmented and virtual reality

Companies are using AI technology and high-level visual recognition tasks such as image classification and object detection to develop AR and virtual reality (VR) hardware and software. Benefits include offering immersive experiences, training and education, helping people with disabilities and providing entertainment. VR and AR have grown remarkably since Ivan Sutherland developed the first VR headset prototype in 1968 to view 3D images. Too heavy to wear, it had to be mounted on the ceiling (Günger and Zengin, 2017[99]). VR companies now provide 360-degree video streaming experiences with much lighter headsets. Pokemon GO drew consumers' attention to AR in 2016 and expectations remain high. AI-embedded applications are already in the marketplace. IKEA provides a mobile app that allows customers to see how a piece of furniture would look and fit in a given space with an accuracy up to 1 millimetre (Jesus, 2018[100]). Some tech companies are developing applications for the visually impaired.[16]

### AI enables interactive AR/VR

As AR/VR develop, AI is being used to help them become interactive, and feature more attractive and intuitive content. AI technologies enable AR/VR to detect user's motion, such as eye movements and hand gestures. This allows it to interpret the motions with high accuracy and customise content in real time according to the user's reaction (Lindell, 2017[101]). For example, AI in VR can detect when a user is looking at a specific visual field and provide full resolution content only in that case. This reduces system resource needs, lags and frame loss (Hall, 2017[102]). Symbiotic development of AR/VR and AI technologies is expected in fields such as marketing research, training simulations and education (Kilpatrick, 2018[103]); (Stanford, 2016[104]).

## *VR for training AI systems*

Some AI systems require large amounts of training data. However, lack of data availability remains an important issue. For example, AI systems in driverless cars must be trained to deal with critical situations, but little actual data about children running onto a street exist. An alternative would be the development of a digital reality. In this case, the AI system would be trained in a computer-simulated environment that faithfully replicates relevant features of the real world. Such a simulated environment can also be used for validating performance of AI systems (e.g. as "driver's license test" for driverless cars) (Slusallek, 2018[105]).

The field of application goes beyond driverless cars. In fact, researchers developed a platform called Household Multimodal Environment (HoME) that would provide a simulated environment to train household robots. HoME has a database with over 45 000 diverse 3D house layouts. It provides a realistic environment for artificial agents to learn through vision, audio, semantics, physics and interaction with objects and other agents (Brodeur et al., 2017[106]).

By allowing AI systems to learn by trial and error, cloud-based VR simulation would be ideal for systems' training, particularly in critical situations. Continuous development in cloud technology would help realise the environment. For example, in October 2017, NVIDIA announced a cloud-based VR simulator that can replicate accurate physics in real-world environments. It is expected that developers will create a new training ground for AI systems within a few years (Solotko, 2017[107]).

## References

Aletras, N. et al. (2016), "Predicting judicial decisions of the European Court of Human Rights: A natural language processing perspective", *PeerJ Computer Science*, Vol. 2, p. e93, http://dx.doi.org/10.7717/peerj-cs.93.  [86]

Angwin, J. et al. (2016), "Machine bias: There's software used across the country to predict future criminals. And it's biased against blacks", *ProPublica*, https://www.propublica.org/article/machine-bias-risk-assessments-in-criminal-sentencing.  [83]

Ashby, M. (2017), "The value of CCTV surveillance cameras as an investigative tool: An empirical analysis", *European Journal on Criminal Policy and Research*, Vol. 23/3, pp. 441-459, http://dx.doi.org/10.1007/s10610-017-9341-6.  [94]

Barocas, S. and A. Selbst (2016), "Big data's disparate impact", *California Law Review*, Vol. 104, pp. 671-729, http://www.californialawreview.org/wp-content/uploads/2016/06/2Barocas-Selbst.pdf.  [74]

Bavitz, C. and K. Hessekiel (2018), *Algorithms and Justice: Examining the Role of the State in the Development and Deployment of Algorithmic Technologies*, Berkman Klein Center for Internet and Society, https://cyber.harvard.edu/story/2018-07/algorithms-and-justice.  [82]

Berg, T. et al. (2018), "On the rise of FinTechs – Credit scoring using digital footprints", *Michael J. Brennan Irish Finance Working Paper Series Research Paper*, No. 18-12, https://papers.ssrn.com/sol3/papers.cfm?abstract_id=3163781.  [32]

Bloom, N. et al. (2017), "Are ideas getting harder to find?", *NBER Working Paper*, No. 23782, http://dx.doi.org/10.3386/w23782.  [52]

Bösch, P. et al. (2018), "Cost-based analysis of autonomous mobility services", *Transport Policy*, Vol. 64, pp. 76-91, https://doi.org/10.1016/j.tranpol.2017.09.005.  [3]

Bose, A. et al. (2016), "The VEICL Act: Safety and security for modern vehicles", *Willamette Law Review*, Vol. 53, p. 137.  [15]

Brayne, S., A. Rosenblat and D. Boyd (2015), "Predictive policing, data & civil rights: A new era of policing and justice", *Pennsylvania Law Review*, Vol. 163/327, http://www.datacivilrights.org/pubs/2015-1027/Predictive_Policing.pdf.  [76]

Brodeur, S. et al. (2017), "HoME: A household multimodal environment", *arXiv* 1107, https://arxiv.org/abs/1711.11017.  [106]

Brodmerkel, S. (2017), "Dynamic pricing: Retailers using artificial intelligence to predict top price you'll pay", *ABC News*, 27 June, http://www.abc.net.au/news/2017-06-27/dynamic-pricing-retailers-using-artificial-intelligence/8638340.  [50]

Brundage, M. et al. (2018), *The Malicious Use of Artificial Intelligence: Forecasting, Prevention, and Mitigation*, Future of Humanity Institute, University of Oxford, Centre for the Study of Existential Risk, University of Cambridge, Centre for a New American Security, Electronic Frontier Foundation and Open AI, https://arxiv.org/ftp/arxiv/papers/1802/1802.07228.pdf.  [108]

Burgess, M. (2018), "UK police are using AI to make custodial decisions but it could be discriminating against the poor", *WIRED*, 1 March, http://www.wired.co.uk/article/police-ai-uk-durham-hart-checkpoint-algorithm-edit.  [79]

Butler, K. et al. (2018), "Machine learning for molecular and materials science", *Nature*, Vol. 559/7715, pp. 547-555, http://dx.doi.org/10.1038/s41586-018-0337-2. [60]

Canadian Institute for Health Information (2013), "Better information for improved health: A vision for health system use of data in Canada", in collaboration with Canada Health Infoway, http://www.cihi.ca/cihi-ext-portal/pdf/internet/hsu_vision_report_en. [63]

Carey, N. and P. Lienert (2018), "Honda to invest $2.75 billion in GM's self-driving car unit", *Reuters*, 3 October, https://www.reuters.com/article/us-gm-autonomous/honda-buys-in-to-gm-cruise-self-driving-unit-idUSKCN1MD1GW. [7]

CFPB (2017), "CFPB explores impact of alternative data on credit access for consumers who are credit invisible", Consumer Financial Protection Bureau, https://www.consumerfinance.gov/about-us/newsroom/cfpb-explores-impact-alternative-data-credit-access-consumers-who-are-credit-invisible/. [37]

Cheng, E. (2017), "Just 10% of trading is regular stock picking, JPMorgan estimates", *CNBC*, 13 June, https://www.cnbc.com/2017/06/13/death-of-the-human-investor-just-10-percent-of-trading-is-regular-stock-picking-jpmorgan-estimates.html. [44]

Chintamaneni, P. (26 June 2017), *How banks can use AI to reduce regulatory compliance burdens*, digitally.cognizant blog, https://digitally.cognizant.com/how-banks-can-use-ai-to-reduce-regulatory-compliance-burdens-codex2710/. [41]

Chow, M. (2017), "AI and machine learning get us one step closer to relevance at scale", *Google*, https://www.thinkwithgoogle.com/marketing-resources/ai-personalized-marketing/. [45]

Christin, A., A. Rosenblat and D. Boyd (2015), *Courts and Predictive Algorithms*, presentation at the "Data & Civil Rights, A New Era of Policing and Justice" conference, Washington, 27 October, http://www.law.nyu.edu/sites/default/files/upload_documents/Angele%20Christin.pdf. [81]

Civardi, C. (2017), *Video Surveillance and Artificial Intelligence: Can A.I. Fill the Growing Gap Between Video Surveillance Usage and Human Resources Availability?*, Balzano Informatik, http://dx.doi.org/10.13140/RG.2.2.13330.66248. [96]

CMU (2015), "Uber, Carnegie Mellon announce strategic partnership and creation of advanced technologies center in Pittsburgh", *Carnegie Mellon University News*, 2 February, https://www.cmu.edu/news/stories/archives/2015/february/uber-partnership.html. [8]

Comfort, S. et al. (2018), "Sorting through the safety data haystack: Using machine learning to identify individual case safety reports in social-digital media", *Drug Safety*, Vol. 41/6, pp. 579-590, https://www.ncbi.nlm.nih.gov/pubmed/29446035. [64]

Cooke, A. (2017), *Digital Earth Australia*, presentation at the "AI: Intelligent Machines, Smart Policies" conference, Paris, 26-27 October, http://www.oecd.org/going-digital/ai-intelligent-machines-smart-policies/conference-agenda/ai-intelligent-machines-smart-policies-cooke.pdf. [22]

De Jesus, A. (2018), "Augmented reality shopping and artificial intelligence – Near-term applications", *Emerj*, 18 December, https://www.techemergence.com/augmented-reality-shopping-and-artificial-intelligence/. [51]

Fagnant, D. and K. Kockelman (2015), "Preparing a nation for autonomous vehicles: Opportunities, barriers and policy recommendations", *Transportation Research A: Policy and Practice*, Vol. 77, pp. 167-181, https://www.sciencedirect.com/science/article/pii/S0. [2]

FAO (2017), "Can artificial intelligence help improve agricultural productivity?", *e-agriculture,* [20]
19 December, http://www.fao.org/e-agriculture/news/can-artificial-intelligence-help-improve-agricultural-productivity.

FAO (2009), *How to Feed the World in 2050*, Food and Agriculture Organization of the United [23]
Nations, Rome,
http://www.fao.org/fileadmin/templates/wsfs/docs/expert_paper/How_to_Feed_the_World_in_2050.pdf.

FDA (2018), *FDA permits marketing of artificial intelligence-based device to detect certain* [66]
*diabetes-related eye problems*, Food and Drug Administration, News Release, 11 April,
https://www.fda.gov/NewsEvents/Newsroom/PressAnnouncements/ucm604357.htm.

Ferguson, A. (2014), "Big Data and Predictive Reasonable Suspicion", *SSRN Electronic Journal*, [75]
http://dx.doi.org/10.2139/ssrn.2394683.

Flores, A., K. Bechtel and C. Lowenkamp (2016), *False positives, false negatives, and false* [84]
*analyses: A rejoinder to "Machine bias: There's software used across the country to predict*
*future criminals. And it's biased against blacks"*, Federal probation, 80.

Fridman, L. (8 October 2018), "Tesla autopilot miles", MIT Human-Centered AI blog, [17]
https://hcai.mit.edu/tesla-autopilot-miles/.

Fridman, L. et al. (2018), "MIT autonomous vehicle technology study: Large-scale deep learning [18]
based analysis of driver behavior and interaction with automation", *arXiv*, Vol. 30/September,
https://arxiv.org/pdf/1711.06976.pdf.

FSB (2017), *Artificial Intelligence and Machine Learning in Financial Services: Market* [31]
*Developments and Financial Stability Implications*, Financial Stability Board, Basel.

FT (2018), "US and China back AI bug-detecting projects", *Financial Times, Cyber Security and* [93]
*Artificial Intelligence,* 28 September, https://www.ft.com/content/64fef986-89d0-11e8-affd-da9960227309.

Gartner (2017), "Gartner's worldwide security spending forecast", Gartner, Press Release, 7 [90]
December, https://www.gartner.com/newsroom/id/3836563.

Gordon, M. and V. Stewart (2017), "CFPB insights on alternative data use on credit scoring", [36]
*Law 360,* 3 May, https://www.law360.com/articles/919094/cfpb-insights-on-alternative-data-use-in-credit-scoring.

Günger, C. and K. Zengin (2017), *A Survey on Augmented Reality Applications using Deep* [99]
*Learning,*
https://www.researchgate.net/publication/322332639_A_Survey_On_Augmented_Reality_Applications_Using_Deep_Learning.

Hall, N. (2017), "8 ways AI makes virtual & augmented reality even more real,", *Topbots,* 13 [102]
May, https://www.topbots.com/8-ways-ai-enables-realistic-virtual-augmented-reality-vr-ar/.

Higgins, T. and C. Dawson (2018), "Waymo orders up to 20,000 Jaguar SUVs for driverless [6]
fleet", *The Wall Street Journal,* 27 March, https://www.wsj.com/articles/waymo-orders-up-to-20-000-jaguar-suvs-for-driverless-fleet-1522159944.

Hinds, R. (2018), *How Natural Language Processing is shaping the Future of Communication*, [46]
MarTechSeries, Marketing Technology Insights, 5 February, https://martechseries.com/mts-insights/guest-authors/how-natural-language-processing-is-shaping-the-future-of-communication/.

Hong, P. (27 August 2017), "Using machine learning to boost click-through rate for your ads", LinkedIn blog, https://www.linkedin.com/pulse/using-machine-learning-boost-click-through-rate-your-ads-tay/. [48]

Horgan, J. (2008), "Against prediction: Profiling, policing, and punishing in an actuarial age – by Bernard E. Harcourt", *Review of Policy Research*, Vol. 25/3, pp. 281-282, http://dx.doi.org/10.1111/j.1541-1338.2008.00328.x. [73]

Hutson, M. (2017), "Artificial intelligence prevails at predicting Supreme Court decisions", *Science Magazine,* 2 May, http://www.sciencemag.org/news/2017/05/artificial-intelligence-prevails-predicting-supreme-court-decisions. [87]

Hu, X. (ed.) (2017), "Human-in-the-loop Bayesian optimization of wearable device parameters", *PLOS ONE*, Vol. 12/9, p. e0184054, http://dx.doi.org/10.1371/journal.pone.0184054. [61]

IHS (2017), "Global defence spending to hit post-Cold War high in 2018", *IHS Markit,* 18 December, https://ihsmarkit.com/research-analysis/global-defence-spending-to-hit-post-cold-war-high-in-2018.html. [89]

Inners, M. and A. Kun (2017), *Beyond Liability: Legal Issues of Human-Machine Interaction for Automated Vehicles*, Proceedings of the 9th International Conference on Automotive User Interfaces and Interactive Vehicular Applications, September, pp. 245-253, http://dx.doi.org/10.1145/3122986.3123005. [14]

ISACA (2016), *The State of Cybersecurity: Implications for 2016*, An ISACA and RSA Conference Survey, Cybersecurity Nexus, https://www.isaca.org/cyber/Documents/state-of-cybersecurity_res_eng_0316.pdf. [91]

ITF (2018), *Safer Roads with Automated Vehicles?*, International Transport Forum, https://www.itf-oecd.org/sites/default/files/docs/safer-roads-automated-vehicles.pdf. [12]

Jagtiani, J. and C. Lemieux (2019), "The roles of alternative data and machine learning in Fintech lending: Evidence from the LendingClub Consumer Platform", *Working Paper*, No. 18-15, Federal Reserve Bank of Philadelphia, http://dx.doi.org/10.21799/frbp.wp.2018.15. [30]

Jenkins, N. (2015), "245 million video surveillance cameras installed globally in 2014", *IHS Markit, Market Insight,* 11 June, https://technology.ihs.com/532501/245-million-video-surveillance-cameras-installed-globally-in-2014. [95]

Jesus, A. (2018), "Augmented reality shopping and artificial intelligence – near-term applications", *Emerj,* 12 December, https://www.techemergence.com/augmented-reality-shopping-and-artificial-intelligence/. [100]

Joh, E. (2017), "The undue influence of surveillance technology companies on policing", *New York University Law Review*, Vol. 91/101, http://dx.doi.org/10.2139/ssrn.2924620. [77]

Jouanjean, M. (2019), "Digital opportunities for trade in the agriculture and food sectors", *OECD Food, Agriculture and Fisheries Papers*, No. 122, OECD Publishing, Paris, https://doi.org/10.1787/91c40e07-en. [24]

Kehl, D., P. Guo and S. Kessler (2017), *Algorithms in the Criminal Justice System: Assessing the Use of Risk Assessment in Sentencing, Responsive Communities Initiative*, Responsive Communities Initiative, Berkman Klein Center for Internet & Society, Harvard Law School. [80]

Kilpatrick, S. (2018), "The rising force of deep learning in VR and AR", *Logik,* 28 March, https://www.logikk.com/articles/deep-learning-in-vr-and-ar/. [103]

King, R. et al. (2004), "Functional genomic hypothesis generation and experimentation by a robot scientist", *Nature*, Vol. 427/6971, pp. 247-252, http://dx.doi.org/10.1038/nature02236. [56]

Kleinberg, J. et al. (2017), "Human decisions and machine predictions", *NBER Working Paper*, No. 23180. [85]

Lee, C., D. Baughman and A. Lee (2017), "Deep learning is effective for classifying normal versus age-related macular degeneration OCT images", *Opthamology Retina*, Vol. 1/4, pp. 322-327. [67]

Lee, T. (2018), "Fully driverless Waymo taxis are due out this year, alarming critics", *Ars Technica,* 1 October, https://arstechnica.com/cars/2018/10/waymo-wont-have-to-prove-its-driverless-taxis-are-safe-before-2018-launch/. [10]

Lindell, T. (2017), "Augmented reality needs AI in order to be effective", *AI Business,* 6 November, https://aibusiness.com/holographic-interfaces-augmented-reality/. [101]

Lippert, J. et al. (2018), "Toyota's vision of autonomous cars is not exactly driverless", *Bloomberg Business Week,* 19 September, https://www.bloomberg.com/news/features/2018-09-19/toyota-s-vision-of-autonomous-cars-is-not-exactly-driverless. [5]

Marr, B. (2018), "How AI and machine learning are transforming law firms and the legal sector", *Forbes,* 23 May, https://www.forbes.com/sites/bernardmarr/2018/05/23/how-ai-and-machine-learning-are-transforming-law-firms-and-the-legal-sector/#7587475832c3. [88]

Marr, B. (2018), "How the UK government uses artificial intelligence to identify welfare and state benefits fraud", *Forbes,* 29 October, https://www.forbes.com/sites/bernardmarr/2018/10/29/how-the-uk-government-uses-artificial-intelligence-to-identify-welfare-and-state-benefits-fraud/#f5283c940cb9. [98]

Mar, V. and H. Soyer (2018), "Artificial intelligence for melanoma diagnosis: How can we deliver on the promise?", *Annals of Oncology*, Vol. 29/8, pp. 1625-1628, http://dx.doi.org/10.1093/annonc/mdy193. [68]

Matchar, E. (2017), "AI plant and animal identification helps us all be citizen scientists", *Smithsonian.com,* 7 June, https://www.smithsonianmag.com/innovation/ai-plant-and-animal-identification-helps-us-all-be-citizen-scientists-180963525/. [55]

Mateescu, A. et al. (2015), *Social Media Surveillance and Law Enforcement, New Era of Criminal Justice and Policing*, Data Civil Rights, http://www.datacivilrights.org/pubs/2015-1027/Social_Media_Surveillance_and_Law_Enforce. [78]

Medhi, S., B. Scholkopf and M. Hirsch (2017), "EnhanceNet: Single image super-resolution through automated texture synthesis", *arXiv* 1612.07919, https://arxiv.org/abs/1612.07919. [97]

MIT (2018), "Cybersecurity's insidious new threat: Workforce stress", *MIT Technology Review,* 7 August, https://www.technologyreview.com/s/611727/cybersecuritys-insidious-new-threat-workforce-stress/. [92]

O'Dwyer, R. (2018), *Algorithms are making the same mistakes assessing credit scores that humans did a century ago*, Quartz, 14 May, https://qz.com/1276781/algorithms-are-making-the-same-mistakes-assessing-credit-scores-that-humans-did-a-century-ago/. [34]

OECD (2018), "Artificial intelligence and machine learning in science", *OECD Science, Technology and Innovation Outlook 2018: Adapting to Technological and Societal Disruption*, No. 5, OECD Publishing, Paris. [57]

OECD (2018), *OECD Science, Technology and Innovation Outlook 2018: Adapting to Technological and Societal Disruption*, OECD Publishing, Paris, https://dx.doi.org/10.1787/sti_in_outlook-2018-en. [53]

OECD (2018), "Personalised pricing in the digital era – Note by the United Kingdom", Key paper for the joint meeting of the OECD Consumer Protection and Competition committees, OECD, Paris, http://www.oecd.org/daf/competition/personalised-pricing-in-the-digital-era.htm. [109]

OECD (2018), *Structural Analysis Database (STAN)*, Rev. 4, Divisions 49 to 53, http://www.oecd.org/industry/ind/stanstructuralanalysisdatabase.htm (accessed on 31 January 2018). [1]

OECD (2017), *New Health Technologies: Managing Access, Value and Sustainability*, OECD Publishing, Paris, http://dx.doi.org/10.1787/9789264266438-en. [69]

OECD (2017), *OECD Digital Economy Outlook 2017*, OECD Publishing, Paris, http://dx.doi.org/10.1787/9789264276284-en. [26]

OECD (2017), *Technology and Innovation in the Insurance Sector*, OECD Publishing, Paris, https://www.oecd.org/finance/Technology-and-innovation-in-the-insurance-sector.pdf (accessed on 28 August 2018). [39]

OECD (2016), *Recommendation of the Council on Health Data Governance*, OECD, Paris, https://legalinstruments.oecd.org/en/instruments/OECD-LEGAL-0433. [71]

OECD (2015), *Data-Driven Innovation: Big Data for Growth and Well-Being*, OECD Publishing, Paris, http://dx.doi.org/10.1787/9789264229358-en. [62]

OECD (2014), "Skills and Jobs in the Internet Economy", *OECD Digital Economy Papers*, No. 242, OECD Publishing, Paris, https://dx.doi.org/10.1787/5jxvbrjm9bns-en. [19]

Ohnsman, A. (2018), "Waymo dramatically expanding autonomous taxi fleet, eyes sales to individuals", *Forbes,* 31 May, https://www.forbes.com/sites/alanohnsman/2018/05/31/waymo-adding-up-to-62000-minivans-to-robot-fleet-may-supply-tech-for-fca-models. [4]

ORAD (2016), "Taxonomy and definitions for terms related to driving automation systems for on-road motor vehicles", On-Road Automated Driving (ORAD) Committee, SAE International, http://dx.doi.org/10.4271/j3016_201609. [9]

Ornstein, C. and K. Thomas (2018), *Sloan Kettering's cozy deal with start-up ignites a new uproar*, 20 September, https://www.nytimes.com/2018/09/20/health/memorial-sloan-kettering-cancer-paige-ai.html. [70]

Patton, E. (2018), *Integrating Artificial Intelligence for Scaling Internet of Things in Health Care*, OECD-GCOA-Cornell-Tech Expert Consultation on Growing and Shaping the Internet of Things Wellness and Care Ecosystem, 4-5 October, New York. [65]

Plummer, L. (2017), "This is how Netflix's top-secret recommendation system works", *WIRED,* 22 August, https://www.wired.co.uk/article/how-do-netflixs-algorithms-work-machine-learning-helps-to-predict-what-viewers-will-like. [47]

Press, G. (2017), "Equifax and SAS leverage AI and deep learning to improve consumer access to credit", *Forbes,* 20 February, https://www.forbes.com/sites/gilpress/2017/02/20/equifax-and-sas-leverage-ai-and-deep-learning-to-improve-consumer-access-to-credit/2/#2ea15ddd7f69. [29]

Rakestraw, R. (2017), "Can artificial intelligence help feed the world?", *Forbes,* 6 September, https://www.forbes.com/sites/themixingbowl/2017/09/05/can-artificial-intelligence-help-feed-the-world/#16bb973646db.    [25]

Roeland, C. (2017), *EC Perspectives on the Earth Observation*, presentation at the "AI: Intelligent Machines, Smart Policies" conference, Paris, 26-27 October, http://www.oecd.org/going-digital/ai-intelligent-machines-smart-policies/conference-agenda/ai-intelligent-machines-smart-policies-roeland.pdf.    [21]

Rollet, C. (2018), "The odd reality of life under China's all-seeing credit score system", *WIRED*, 5 June, https://www.wired.co.uk/article/china-blacklist.    [35]

Segler, M., M. Preuss and M. Waller (2018), "Planning chemical syntheses with deep neural networks and symbolic AI", *Nature*, Vol. 555/7698, pp. 604-610, http://dx.doi.org/10.1038/nature25978.    [59]

Simon, M. (2017), "Phone-powered AI spots sick plants with remarkable accuracy", *WIRED,* 2 February, https://www.wired.com/story/plant-ai/.    [28]

Slusallek, P. (2018), *Artificial Intelligence and Digital Reality: Do We Need a CERN for AI?*, The Forum Network, OECD, Paris, https://www.oecd-forum.org/channels/722-digitalisation/posts/28452-artificial-intelligence-and-digital-reality-do-we-need-a-cern-for-ai.    [105]

Sohangir, S. et al. (2018), "Big data: Deep learning for financial sentiment analysis", *Journal of Big Data*, Vol. 5/1, http://dx.doi.org/10.1186/s40537-017-0111-6.    [40]

Sokolin, L. and M. Low (2018), *Machine Intelligence and Augmented Finance: How Artificial Intelligence Creates $1 Trillion Dollar of Change in the Front, Middle and Back Office*, Autonomous Research LLP, https://next.autonomous.com/augmented-finance-machine-intelligence.    [38]

Solotko, S. (2017), "Virtual reality is the next training ground for artificial intelligence", *Forbes,* 11 October, https://www.forbes.com/sites/tiriasresearch/2017/10/11/virtual-reality-is-the-next-training-ground-for-artificial-intelligence/#6e0c59cc57a5.    [107]

Song, H. (2017), "JPMorgan software does in seconds what took lawyers 360,000 hours", *Bloomberg.com,* 28 February, https://www.bloomberg.com/news/articles/2017-02-28/jpmorgan-marshals-an-army-of-developers-to-automate-high-finance.    [42]

Spangler, S. et al. (2014), *Automated Hypothesis Generation based on Mining Scientific Literature*, ACM Press, New York, http://dx.doi.org/10.1145/2623330.2623667.    [54]

Stanford (2016), *Artificial Intelligence and Life in 2030*, AI100 Standing Committee and Study Panel, Stanford University, https://ai100.stanford.edu/2016-report.    [104]

Surakitbanharn, C. et al. (2018), *Preliminary Ethical, Legal and Social Implications of Connected and Autonomous Transportation Vehicles (CATV)*, Purdue University, https://www.purdue.edu/discoverypark/ppri/docs/Literature%20Review_CATV.pdf.    [16]

Voegeli, V. (2016), "Credit Suisse, CIA-funded palantir to target rogue bankers", *Bloomberg,* 22 March, https://www.bloomberg.com/news/articles/2016-03-22/credit-suisse-cia-funded-palantir-build-joint-compliance-firm.    [43]

Waid, B. (2018), "AI-enabled personalization: The new frontier in dynamic pricing", *Forbes,* 9 July, https://www.forbes.com/sites/forbestechcouncil/2018/07/09/ai-enabled-personalization-the-new-frontier-in-dynamic-pricing/#71e470b86c1b.    [49]

Walker-Smith, B. (2013), "Automated vehicles are probably legal in the United States", *Texas A&M Law Review*, Vol. 1/3, pp. 411-521.                                                                                      [11]

Webb, L. (2017), *Machine Learning in Action*, presentation at the "AI: Intelligent Machines, Smart Policies" conference, Paris, 26-27 October, http://www.oecd.org/going-digital/ai-intelligent-machines-smart-policies/conference-agenda/ai-intelligent-machines-smart-policies-webb.pdf.                                                                     [27]

Welsch, D. and E. Behrmann (2018), "Who's winning the self-driving car race?", *Bloomberg*, 7 May, https://www.bloomberg.com/news/features/2018-05-07/who-s-winning-the-self-driving-car-race.                                                                     [13]

Williams, K. et al. (2015), "Cheaper faster drug development validated by the repositioning of drugs against neglected tropical diseases", *Journal of The Royal Society Interface*, Vol. 12/104, pp. 20141289-20141289, http://dx.doi.org/10.1098/rsif.2014.1289.           [58]

Wyllie, D. (2013), "How 'big data' is helping law enforcement", *PoliceOne.Com*, 20 August, https://www.policeone.com/police-products/software/Data-Information-Sharing-Software/articles/6396543-How-Big-Data-is-helping-law-enforcement/.                         [72]

Zeng, M. (2018), "Alibaba and the future of business", *Harvard Business Review*, September-October, https://hbr.org/2018/09/alibaba-and-the-future-of-business.                         [33]

Notes

[1] STAN Industrial Analysis, 2018, value added of "Transportation and Storage" services, ISIC Rev. 4 Divisions 49 to 53, as a share of total value added, 2016 unweighted OECD average. The 2016 weighted OECD average was 4.3%.

[2] From https://www.crunchbase.com/.

[3] In 1989, Fair, Isaac and Company (FICO) introduced the FICO credit score. It is still used by the majority of banks and credit grantors.

[4] The OECD Committee on Consumer Protection, adopted the definition of personalised pricing given by the United Kingdom's Office of Fair Trading: "Personalised pricing can be defined as a form of price discrimination, in which: 'businesses may use information that is observed, volunteered, inferred, or collected about individuals' conduct or characteristics, to set different prices to different consumers (whether on an individual or group basis), based on what the business thinks they are willing to pay" (OECD, 2018[109]). If utilised by vendors, personalised pricing could result in some consumers paying less for a given good or service, while others pay more than they would have done if all consumers were offered the same price.

[5] This section draws on work by the OECD Committee for Scientific and Technological Policy, particularly Chapter 3 – "Artificial Intelligence and Machine Learning in Science" – of OECD (2018[53]). The main authors of that chapter were Professor Stephen Roberts, of Oxford University, and Professor Ross King, of Manchester University.

[6] See https://iris.ai/.

[7] See https://www.atomwise.com/2015/03/24/new-ebola-treatment-using-artificial-intelligence/.

[8] See https://www.bbc.com/news/technology-40592520.

[9] See https://www.nytimes.com/2018/09/20/health/memorial-sloan-kettering-cancer-paige-ai.html.

[10] *State of Wisconsin v. Loomis*, 881 N.W.2d 749 (Wis. 2016).

[11] *Loomis v. Wisconsin*, 137 S.Ct. 2290 (2017).

[12] While the importance of public spending on AI technologies for defence purposes is acknowledged, this area of study falls outside the scope of this publication.

[13] Unless otherwise specified, throughout this work "digital security" refers to the management of economic and social risks resulting from breaches of availability, integrity and confidentiality of information and communication technologies, and data.

[14] A phishing attack is a fraudulent attempt to obtain sensitive information from a target by deceiving it through electronic communications with an illegitimately trustworthy facade. A more labour-intensive modality is spear phishing, which is customised to the specific individual or organisation being targeted by collecting and using sensitive information such as name, gender, affiliation, etc. (Brundage et al., 2018[108]). Spear phishing is the most common infection vector: in 2017, 71% of cyberattacks began with spear phishing emails.

[15] This has been confirmed by the E-Leaders group, which is part of the OECD Public Governance Committee. Its Thematic Group on Emerging Technologies – made up of representatives from 16 countries – focuses mainly on AI and blockchain.

[16] BlindTool (https://play.google.com/store/apps/details?id=the.blindtool&hl=en) and Seeing AI (https://www.microsoft.com/en-us/seeing-ai) are examples of the applications.

# 4. Public policy considerations

*This chapter explores public policy considerations to ensure that artificial intelligence (AI) systems are trustworthy and human-centred. It covers concerns related to ethics and fairness; the respect of human democratic values, including privacy; and the dangers of transferring existing biases from the analogue world into the digital world, including those related to gender and race. The need to progress towards more robust, safe, secure and transparent AI systems with clear accountability mechanisms for their outcomes is underlined.*

*Policies that promote trustworthy AI systems include those that encourage investment in responsible AI research and development; enable a digital ecosystem where privacy is not compromised by a broader access to data; enable small and medium-sized enterprises to thrive; support competition, while safeguarding intellectual property; and facilitate transitions as jobs evolve and workers move from one job to the next.*

Human-centred AI

Artificial intelligence (AI) plays an increasingly influential role. As the technology diffuses, the potential impacts of its predictions, recommendations or decisions on people's lives increase as well. The technical, business and policy communities are actively exploring how best to make AI human-centred and trustworthy, maximise benefits, minimise risks and promote social acceptance.

---

**Box 4.1. "Black box" AI systems present new challenges
from previous technological advancements**

Neural networks have often been referred to as a "black box". Although the behaviour of such systems can indeed be monitored, the term "black box" reflects the considerable difference between the ability to monitor previous technologies compared to neural networks. Neural networks iterate on the data they are trained on. They find complex, multi-variable probabilistic correlations that become part of the model that they build. However, they do not indicate how data could interrelate (Weinberger, 2018[1]). The data are far too complex for the human mind to understand. Characteristics of AI that differ from previous technological advancements and affect transparency and accountability include:

- **Discoverability**: Rules-based algorithms can be read and audited rule-by-rule, making it comparatively straightforward to find certain types of errors. By contrast, certain types of machine-learning (ML) systems, notably neural networks, are simply abstract mathematical relationships between factors. These can be extremely complex and difficult to understand, even for those who program and train them (OECD, 2016).

- **Evolving nature**: Some ML systems iterate and evolve over time and may even change their own behaviour in unforeseen ways.

- **Not easily repeatable**: A specific prediction or decision may only appear when the ML system is presented with specific conditions and data, which are not necessarily repeatable.

- **Increased tensions in protecting personal and sensitive data**:

   o **Inferences**: Even in the absence of protected or sensitive data, AI systems may be able to infer these data and correlations from proxy variables that are not personal or sensitive, such as purchasing history or location (Kosinski, Stillwell and Graepel, 2013[2]).

   o **Improper proxy variables**: Policy and technical approaches to privacy and non-discrimination have tended to minimise data collected, prohibit use of certain data or remove data to prevent their use. But an AI system might base a prediction on proxy data that bear a close relationship to the forbidden and non-collected data. Furthermore, the only way to detect these proxies is to also collect sensitive or personal data such as race. If such data are collected, then it becomes important to ensure they are only used in appropriate ways.

   o **The data-privacy paradox**: For many AI systems, more training data can improve the accuracy of AI predictions and help reduce risk of bias from skewed samples. However, the more data collected, the greater the privacy risks to those whose data are collected.

---

Some types of AI – often referred to as "black boxes" – raise new challenges compared to previous technological advancements (Box 4.1). Given these challenges, the OECD – building on the work of the AI Group of Experts at the OECD (AIGO) – identified key priorities for human-centred AI. First, it should contribute to inclusive and sustainable growth and well-being. Second, it should respect human-centred values and fairness. Third, the use of AI and how its systems operate should be transparent. Fourth, AI systems should be robust and safe. Fifth, there should be accountability for the results of AI predictions and the ensuing decisions. Such measures are viewed as critical for high-stakes' predictions. They are also important in business recommendations or for less impactful uses of AI.

## Inclusive and sustainable growth and well-being

### *AI holds significant potential to advance the agenda towards meeting the Sustainable Development Goals*

AI can be leveraged for social good and to advance meeting the United Nations Sustainable Development Goals (SDGs) in areas such as education, health, transport, agriculture and sustainable cities, among others. Many public and private organisations, including the World Bank, a number of United Nations agencies and the OECD, are working to leverage AI to help advance the SDGs.

### *Ensuring that AI development is equitable and inclusive is a growing priority*

Ensuring that AI development is equitable and inclusive is a growing priority. This is especially true in light of concerns about AI exacerbating inequality or increasing existing divides within and between developed and developing countries. These divides exist due to concentration of AI resources – AI technology, skills, datasets and computing power – in a few companies and nations. In addition, there is concern that AI could perpetuate biases (Talbot et al., 2017[3]). Some fear AI could have a disparate impact on vulnerable and under-represented populations. These include the less educated, low skilled, women and elderly, particularly in low- and middle-income countries (Smith and Neupane, 2018[4]). Canada's International Development Research Centre recently recommended the formation of a global AI for Development fund. This would establish AI Centres of Excellence in low- and middle-income countries to support the design and implementation of evidence-based inclusive policy (Smith and Neupane, 2018[4]). Its goal is to ensure that AI benefits are well distributed and lead to more egalitarian societies. Inclusive AI initiatives aim to ensure that economic gains from AI in societies are widely shared and that no one is left behind.

Inclusive and sustainable AI is an area of focus for countries such as India (NITI, 2018[5]); companies such as Microsoft;[1] and academic groups such as the Berkman Klein Center at Harvard. For example, Microsoft has launched projects such as the Seeing AI mobile application, which helps the visually impaired. It scans and recognises all the elements surrounding a person and provides an oral description. Microsoft is also investing USD 2 million in qualified initiatives to leverage AI to tackle sustainability challenges such as biodiversity and climate change (Heiner and Nguyen, 2018[6]).

### Human rights and ethical codes

#### International human rights law embodies ethical norms

International human rights law embodies ethical norms. AI can support the fulfilment of human rights, as well as create new risks that human rights might be deliberately or accidently violated. Human rights law, together with the legal and other institutional structures related to it, could also serve as one of the tools to help ensure human-centred AI (Box 4.2).

---

**Box 4.2. Human rights and AI**

International human rights refer to a body of international laws, including the International Bill of Rights,[1] as well as regional human rights systems developed over the past 70 years around the world. Human rights provide a set of universal minimum standards based on, among others, values of human dignity, autonomy and equality, in line with the rule of law. These standards and the legal mechanisms linked to them create legally enforceable obligations for countries to respect, protect and fulfil human rights. They also require that those whose rights have been denied or violated be able to obtain effective remedy.

Specific human rights include the right to equality, the right to non-discrimination, the right to freedom of association, the right to privacy and economic, social and cultural rights such as the right to education or the right to health.

Recent intergovernmental instruments such as the United Nations *Guiding Principles on Business and Human Rights* (OHCHR, 2011[7]) have also addressed private actors in the context of human rights. They provide for a "responsibility" on private actors to respect human rights. In addition, the 2011 update of government-backed recommendations to business in the *OECD Guidelines for Multinational Enterprises* (OECD, 2011[8]) contains a chapter on human rights.

Human rights overlap with wider ethical concerns and with other areas of regulation relevant to AI, such as personal data protection or product safety law. However, these other concerns and issues often have different scope.

1. Comprised of the Universal Declaration of Human Rights, the International Covenant on Civil and Political Rights and the International Covenant on Economic, Social and Cultural Rights.

---

#### AI promises to advance human rights

Given the potential breadth of its application and use, AI promises to advance the protection and fulfilment of human rights. Examples include using AI in the analysis of patterns in food scarcity to combat hunger, improving medical diagnosis and treatment or making health services more widely available and accessible, and shedding light on discrimination.

#### AI could also challenge human rights

AI may also pose a number of human rights challenges that are often reflected in discussions on AI and ethics more broadly. Specific AI systems could violate, or be used to violate, human rights accidentally or deliberately. Much focus is placed on accidental impacts. ML algorithms that predict recidivism, for example, may have undetected bias. Yet AI technologies can also be linked to intentional violations of human rights. Examples include the use of AI technologies to find political dissidents, and restricting individuals' rights to freedom of expression or to participate in political life. In these cases, the violation in itself is usually not unique to the use of AI. However, it could be exacerbated by AI's sophistication and efficiency.

The use of AI may also pose unique challenges in situations where human rights impacts are unintentional or difficult to detect. The reason can be the use of poor-quality training data, system design or complex interactions between the AI system and its environment. One example is algorithmic exacerbation of hate speech or incitement to violence on line. Another example is the unintentional amplifying of fake news, which could impact the right to take part in political and public affairs. The likely scale and impact of harm will be linked to the scale and potential impact of decisions by any specific AI system. For example, a decision by a news recommendation AI system has a narrower potential impact than a decision by an algorithm predicting the risk of recidivism of parole inmates.

### *Human rights frameworks complemented by AI ethical codes*

Ethical codes can address the risk that AI might not operate in a human-centred manner or align with human values. Both private companies and governments have adopted a large number of ethical codes relating to AI.

For example, Google-owned DeepMind also created a DeepMind Ethics & Society unit in October 2017.[2] The unit's goal is to help technologists understand the ethical implications of their work and help society decide how AI can be beneficial. The unit will also fund external research on algorithmic bias, the future of work, lethal autonomous weapons and more. Google itself announced a set of ethical principles to guide its research, product development and business decisions.[3] It published a white paper on AI governance, identifying issues for further clarification with governments and civil societies.[4] Microsoft's AI vision is to "amplify human ingenuity with intelligent technology" (Heiner and Nguyen, 2018[6]). The company has launched projects to ensure inclusive and sustainable development.

Human rights law, together with its institutional mechanisms and wider architecture, provides the direction and basis to ensure the ethical and human-centred development and use of AI in society.

### *Leveraging human rights frameworks in the AI context presents advantages*

The advantages of leveraging human rights frameworks in the AI context include established institutions, jurisprudence, universal language and international acceptance:

- **Established institutions**: A wide international, regional and national human rights infrastructure has been developed over time. It is comprised of intergovernmental organisations, courts, non-governmental organisations, academia, and other institutions and communities where human rights can be invoked and remedy sought.

- **Jurisprudence**: As legal norms, the values protected by human rights are operationalised and made concrete, and legally binding, in specific situations through jurisprudence and the interpretative work by international, regional and national institutions.

- **Universal language**: Human rights provide a universal language for a global issue. This, together with the human rights infrastructure, can help enfranchise a wider variety of stakeholders. They can thus participate in the debate on the place of AI in society, alongside the AI actors directly involved in the AI lifecycle.

- **International acceptance**: Human rights have broad international acceptance and legitimacy. The mere perception that an actor may violate human rights can be significant, since the associated reputational costs can be high.

*A human rights approach to AI can help identify risks, priorities, vulnerable groups and provide remedy*

- **Risk identification**: Human rights frameworks can help identify risks of harm. In particular, they can carry out human rights due diligence such as human rights impact assessments (HRIAs) (Box 4.3).

- **Core requirements**: As minimum standards, human rights define inviolable core requirements. For example, in the regulation of expression on social networks, human rights jurisprudence helps demarcate hate speech as a red line.

- **Identifying high-risk contexts**: Human rights can be a useful tool to identify high-risk contexts or activities. In such situations, increased care is needed or AI could be deemed unfit for use.

- **Identifying vulnerable groups or communities**: Human rights can help identify vulnerable or at-risk groups or communities in relation to AI. Some individuals or communities may be under-represented due, for example, to limited smartphone use.

- **Remedy**: As legal norms with attendant obligations, human rights can provide remedy to those whose rights are violated. Examples of remedies include cessation of activity, development of new processes or policies, an apology or monetary compensation.

---

**Box 4.3. Human rights impact assessments**

HRIAs can help determine risks that AI lifecycle actors might not otherwise envisage. To that end, they focus on incidental human impacts rather than optimisation of the technology or its outputs. HRIAs or similar processes could ensure by-design respect for human rights throughout the lifecycle of the technology.

HRIAs assess technology against a wide range of possible human rights impacts, a broad-sweeping approach that is resource-intensive. It can be easier to start with the AI system in question and work outwards. In this way, AI focuses on a limited range of areas where rights challenges appear most likely. Industry organisations can help conduct HRIAs for small and medium-sized enterprises (SMEs) or non-tech companies that deploy AI systems but may not be literate in the technology. The Global Network Initiative exemplifies such an organisation with respect to freedom of expression and privacy. It helps companies plan ahead and incorporate human rights assessments into their plans for new products (https://globalnetworkinitiative.org/).

HRIAs have the drawback of generally being conducted company by company. Conversely, AI systems might involve many actors, which means looking at only one part may be ineffective. Microsoft was the first large technology company to conduct an HRIA on AI in 2018.

---

There are also significant challenges to implement a human rights approach to AI. These are related to how human rights are directed towards countries, how enforcement is tied to jurisdictions, how they are better suited to remediate substantial harms to a small number of individuals and how they can be costly to business:

- **Human rights are directed towards countries, not private actors**. Yet private-sector actors play a key role in AI research, development and deployment. This challenge is not unique to AI. Several intergovernmental initiatives seek to overcome the public/private divide. Beyond such efforts, there is growing recognition that a good human rights record is good for business.[5]

- **Enforcement of human rights is tied to jurisdictions**. Generally, claimants must demonstrate legal standing in a specific jurisdiction. These approaches may not be optimal when cases involve large multinational enterprises and AI systems that span multiple jurisdictions.

- **Human rights are better suited to remediate substantial harms to a small number of individuals**, as opposed to less significant harms suffered by many. In addition, human rights and their structures can seem opaque to outsiders.

- **In some contexts, human rights have a reputation as being costly to business**. Therefore, approaches that put forward ethics, consumer protection or responsible business conduct, as well as the business case for respecting human rights, seem promising.

Some general challenges of AI such as transparency and explainability also apply with respect to human rights (Section "Transparency and explainability"). Without transparency, identifying when human rights have been violated or substantiating a claim of violation is difficult. The same is true for seeking remedy, determining causality and accountability.

## *Personal data protection*

### *AI challenges notions of "personal data" and consent*

AI can increasingly link different datasets and match different types of information with profound consequences. Data held separately were once considered non-personal (or were stripped of personal identifiers, i.e. "de-identified"). With AI, however, non-personal data can be correlated with other data and matched to specific individuals, becoming personal (or "re-identified"). Thus, algorithmic correlation weakens the distinction between personal data and other data. Non-personal data can increasingly be used to re-identify individuals or infer sensitive information about them, beyond what was originally and knowingly disclosed (Cellarius, 2017[9]). In 2007, for example, researchers had already used reportedly anonymous data to link Netflix's list of movie rentals with reviews posted on IMDB. In this way, they identified individual renters and accessed their complete rental history. With more data collected, and technological improvements, such links are increasingly possible. It becomes difficult to assess which data can be considered and will remain non-personal.

It is increasingly difficult to distinguish between sensitive and non-sensitive data in, for example, the European Union's General Data Protection Regulation (GDPR). Some algorithms can infer sensitive information from "non-sensitive" data, such as assessing individuals' emotional state based on their keyboard typing pattern (Privacy International and Article 19, 2018[10]). The use of AI to identify or re-identify data that originally were non-personal or de-identified also presents a legal issue. Protection frameworks, like the OECD Recommendation of the Council concerning *Guidelines Governing the Protection of Privacy and Transborder Flows of Personal Data* ("Privacy Guidelines"), apply to personal data (Box 4.4). Therefore, it is not clear if, or at what point, they apply to data that under some circumstances would be, or could be, identifiable (Office of the Victorian Information

Commissioner, 2018[11]). An extreme interpretation could result in vastly broadening the scope of privacy protection, which would make it difficult to apply.

---

**Box 4.4. The OECD Privacy Guidelines**

The Recommendation of the Council concerning *Guidelines Governing the Protection of Privacy and Transborder Flows of Personal Data* ("Privacy Guidelines") was adopted in 1980 and updated in 2013 (OECD, 2013[12]). It contains definitions of relevant terms, notably defining "personal data" as "any information relating to an identified or identifiable individual (data subject)". It also defines principles to apply when processing personal data. These principles relate to collection limitation (including, where appropriate, consent as means to ensure this principle), data quality, purpose specification, use limitation, security safeguards, openness, individual participation and accountability. They also provide that in implementing the Privacy Guidelines, members should ensure there is no unfair discrimination against data subjects. The implementation of the Privacy Guidelines was to be reviewed in 2019 to take account, among others, of recent developments, including in the area of AI.

---

*AI also challenges personal data protection principles of collection limitation, use limitation and purpose specification*

To train and optimise AI systems, ML algorithms require vast quantities of data. This creates an incentive to maximise, rather than minimise, data collection. With the growth in use of AI devices, and the Internet of Things (IoT), more data are gathered, more frequently and more easily. They are linked to other data, sometimes with little or no awareness or consent on the part of the data subjects concerned.

The patterns identified and evolution of the "learning" are difficult to anticipate. Therefore, the collection and use of data can extend beyond what was originally known, disclosed and consented to by a data subject (Privacy International and Article 19, 2018[10]). This is potentially incompatible with the Privacy Guidelines' principles of collection limitation, use limitation and purpose specification (Cellarius, 2017[9]). These first two principles rely in part on the data subject's consent (as appropriate, recognising that consent may not be feasible in some cases). This consent is either the basis for the collection of personal data, or for its use for other purposes than originally specified. AI technologies such as deep learning that are difficult to understand or monitor are also difficult to explain to the data subjects concerned. This is a challenge for companies. They report the exponential rate at which AI gains access to, analyses and uses data is difficult to reconcile with these data protection principles (OECD, 2018[13]).

The combination of AI technologies with developments in the IoT, i.e. the connection of an increasing number of devices and objects over time to the Internet, exacerbates the challenges. The increasing combination of AI and IoT technologies (e.g. IoT devices equipped with AI, or AI algorithms used to analyse IoT data) means that more data, including personal data, are constantly gathered. These can be increasingly linked and analysed. On the one hand, there is an increased presence of devices collecting information (e.g. surveillance cameras or autonomous vehicles [AVs]). On the other, there is better AI technology (e.g. facial recognition). The combination of these two trends risks leading to more invasive outcomes than either factor separately (Office of the Victorian Information Commissioner, 2018[11]).

*AI can also empower individual participation and consent*

AI carries potential to enhance personal data. For example, initiatives to build AI systems around principles of privacy by design and privacy by default are ongoing within a number of technical standards organisations. For the most part, they use and adapt privacy guidelines, including the OECD Privacy Guidelines. Additionally, AI is used to offer individuals tailored personalised services based on their personal privacy preferences, as learned over time (Office of the Victorian Information Commissioner, 2018[11]). These services can help individuals navigate between the different personal data processing policies of different services and ensure their preferences are considered across the board. In so doing, AI empowers meaningful consent and individual participation. A team of researchers, for example, developed Polisis, an automated framework that uses neural network classifiers to analyse privacy policies (Harkous, 2018[14]).

## *Fairness and ethics*

### *ML algorithms can reflect the biases implicit in their training data*

To date, AI policy initiatives feature ethics, fairness and/or justice prominently. There is significant concern that ML algorithms tend to reflect and repeat the biases implicit in their training data, such as racial biases and stereotyped associations. Because technological artefacts often embody societal values, discussions of fairness should articulate which societies technologies should serve, who should be protected and with what core values (Flanagan, Howe and Nissenbaum, 2008[15]). Disciplines such as philosophy, law and economy have grappled with different notions of fairness for decades from several angles. They illustrate the broad range of possible visions of fairness and the implications for policy.

### *Philosophical, legal and computational notions of fairness and ethical AI vary*

Philosophy focuses on concepts of right and wrong conduct, good and evil, and morality. Three major philosophical theories are relevant in the context of ethical AI (Abrams et al., 2017[16]):

- **The fundamental human rights approach**, associated with Immanuel Kant, identifies the formal principles of ethics, which are specific rights such as privacy or freedom. It protects these principles by regulation, which AI systems should respect.

- **The utilitarian approach**, pursued by Jeremy Bentham and John Stuart Mill, focuses on public policies that maximise human welfare based on economic cost benefit analyses. For AI, the utilitarian approach raises the question of *whose* welfare to maximise (e.g. individuals, family, society or institutions/governments), which may impact algorithm design.

- **The virtue ethics approach**, based on Aristotle's work, focuses on the values and ethical norms needed for a society to support people in their everyday efforts to live a life worth living. This raises the question of which values and which ethical norms warrant protection.

The law often uses the terms "equality" and "justice" to represent concepts of fairness. The two major legal approaches to fairness are individual fairness and group fairness.

- **Individual fairness** represents the notion of equality before the law. It implies that everyone should be treated equally and not discriminated against in view of special attributes. Equality is recognised as an international human right.

- **Group fairness** focuses on the fairness of the outcome. It ensure the outcome does not differ in any systematic manner for people who, based on a protected characteristic (e.g. race or gender), belong to different groups. It reasons that differences and historical circumstances may lead different groups to react to situations differently. Different countries' approaches to group fairness differ significantly. Some, for example, use positive discrimination.

AI system designers have been considering how to represent fairness in AI systems. Different definitions of fairness embody different possible approaches (Narayanan, 2018[17]):

- The "**unaware approach**", whereby an AI system should be unaware of any identifiable factors, accompanies the individual fairness legal approach. In this case, the AI system does not consider data on sensitive or prohibited attributes, such as gender, race and sexual orientation (Yona, 2017[18]). However, many other factors may be correlated with the protected/prohibited attribute (such as gender). Removing them could limit the accuracy of an AI system.

- **Fairness through awareness** acknowledges group differences and aims to treat similar individuals in the same way. The challenge is, however, to determine who should be treated similarly to whom. Understanding who should be considered similar for a particular task requires knowledge of sensitive attributes.

- **Group fairness approaches** focus on ensuring that outcomes do not differ systematically for people who belong to different groups. There is concern about the potential for AI systems to be unfair, perpetuating or reinforcing traditional biases, since they often rely on data sets representing past activity.

Different notions of fairness translate into different results for different groups in society and different types of stakeholders. They cannot all be simultaneously achieved. Policy considerations and in some cases, choices, should inform technological design choices that could adversely impact specific groups.

### *AI in human resources illustrates the opportunity and challenge of AI for bias*

In human resources, use of AI either perpetuates bias in hiring or helps uncover and reduce harmful bias. A Carnegie Mellon study exploring patterns of online job postings showed that an ad for higher-paid executives was displayed 1 816 times to men and just 311 times to women (Simonite, 2018[19]). Thus, one potential area for human-AI collaboration is to ensure that AI applications for hiring and evaluation are transparent. They should not codify biases, e.g. by automatically disqualifying diverse candidates for roles in historically non-diverse settings (OECD, 2017[20]).

### *Several approaches can help mitigate discrimination in AI systems*

Approaches proposed to mitigate discrimination in AI systems include awareness building; organisational diversity policies and practices; standards; technical solutions to detect and correct algorithmic bias; and self-regulatory or regulatory approaches. For example, in predictive policing systems, some propose algorithmic impact assessments or statements. These would require police departments to evaluate the efficacy, benefits and potential discriminatory effects of all available choices for predictive policing technologies (Selbst, 2017[21]). Accountability and transparency are important to achieve fairness. However, even combined they do not guarantee it (Weinberger, 2018[22]); (Narayanan, 2018[17]).

*Striving for fairness in AI systems may call for trade-offs*

AI systems are expected to be "fair". This aims to result in, for example, only the riskiest defendants remaining in jail or the most suitable lending plan being proposed based on ability to pay. **False positive errors** indicate a misclassification of a person or a behaviour. For example, they could wrongly predict a defendant will reoffend when he or she will not. They could also wrongly predict a disease that is not there. **False negative errors** represent cases in which an AI system wrongly predicts, for example, that a defendant will not reoffend. As another example, a test may wrongly indicate the absence of a disease.

Group fairness approaches acknowledge different starting points for different groups. They try to account for differences mathematically by ensuring "equal accuracy" or equal error rates across all groups. For example, they would wrongly classify the same percentage of men and women as reoffenders (or equalise the false positives and false negatives).

Equalising false positives and false negatives creates a challenge. False negatives are often viewed as more undesirable and risky than false positives because they are more costly (Berk and Hyatt, 2015[23]). For example, the cost to a bank of lending to someone an AI system predicted would not default – but who does default – is greater than the gain from that loan. Someone diagnosed as not having a disease who does have that disease may suffer significantly. Equalising true positives and true negatives can also lead to undesirable outcomes. They could, for example, incarcerate women who pose no safety risk so that the same proportion of men and women are released (Berk and Hyatt, 2015[23]). Some approaches aim, for example, to equalise both false positives and false negatives at the same time. However, it is difficult to simultaneously satisfy different notions of fairness (Chouldechova, 2016[24]).

*Policy makers could consider the appropriate treatment of sensitive data in the AI context*

The appropriate treatment of sensitive data could be reconsidered. In some cases, organisations may need to maintain and use sensitive data to ensure their algorithms do not inadvertently reconstruct this data. Another priority for policy is to monitor unintended feedback loops. When police go to algorithmically identified "high crime" areas, for example, this could lead to distorted data collection. It would further bias the algorithm – and society – against these neighbourhoods (O'Neil, 2016[25]).

## Transparency and explainability

### *Transparency about the use of AI and as to how AI systems operate is key*

The technical and the policy meanings of the term "transparency" differ. For policy makers, transparency traditionally focuses on how a decision is made, who participates in the process and the factors used to make the decision (Kosack and Fung, 2014[26]). From this perspective, transparency measures might disclose how AI is being used in a prediction, recommendation or decision. They might also disclose when a user is interacting with an AI-powered agent.

For technologists, transparency of an AI system focuses largely on process issues. It means allowing people to understand how an AI system is developed, trained and deployed. It may also include insight into factors that impact a specific prediction or decision. It does not usually include sharing specific code or datasets. In many cases, the systems are too complex for these elements to provide meaningful transparency (Wachter, Mittelstadt and Russell, 2017[27]). Moreover, sharing specific code or datasets could reveal trade secrets or disclose sensitive user data.

More generally, awareness and understanding of AI reasoning processes is viewed as important for AI to become commonly accepted and useful.

### Approaches to transparency in AI systems

Experts at Harvard University in the Berkman Klein Center Working Group on Explanation and the Law identify approaches to improve transparency of AI systems, and note that each entails trade-offs (Doshi-Velez et al., 2017[28]). An additional approach is that of optimisation transparency, i.e. transparency about the goals of an AI system and about their results. These approaches are: i) theoretical guarantees; ii) empirical evidence; and iii) explanation (Table 4.1).

**Table 4.1. Approaches to improve the transparency and accountability of AI systems**

| Approach | Description | Well-suited contexts | Poorly suited contexts |
|---|---|---|---|
| Theoretical guarantees | In some situations, it is possible to give theoretical guarantees about an AI system backed by proof. | The environment is fully observable (e.g. the game of Go) and both the problem and solution can be formalised. | The situation cannot be clearly specified (most real-world settings). |
| Statistical evidence/ probability | Empirical evidence measures a system's overall performance, demonstrating the value or harm of the system, without explaining specific decisions. | Outcomes can be fully formalised; it is acceptable to wait to see negative outcomes to measure them; issues may only be visible in aggregate. | The objective cannot be fully formalised; blame or innocence can be assigned for a particular decision. |
| Explanation | Humans can interpret information about the logic by which a system took a particular set of inputs and reached a particular conclusion. | Problems are incompletely specified, objectives are not clear and inputs could be erroneous. | Other forms of accountability are possible. |

*Source*: adapted from Doshi-Velez et al. (2017[28]), "Accountability of AI under the law: The role of explanation", https://arxiv.org/pdf/1711.01134.pdf.

### Some systems offer theoretical guarantees of their operating constraints

In some cases, **theoretical guarantees** can be provided, meaning the system will demonstrably operate within narrow constraints. Theoretical guarantees apply to situations in which the environment is fully observable and both the problem and the solution can be fully formalised, such as in the game of Go. In such cases, certain kinds of outcomes cannot happen, even if an AI system processes new kinds of data. For example, a system could be designed to provably follow agreed-upon processes for voting and vote counting. In this case, explanation or evidence may not be required: the system does not need to explain how it reached an outcome because the types of outcomes that cause concern are mathematically impossible. An assessment can be made at an early stage of whether these constrains are sufficient.

### Statistical evidence of overall performance can be provided in some cases

In some cases, relying on **statistical evidence** of a system's overall performance may be sufficient. Evidence that an AI system measurably increases a specific societal or individual value or harm may be sufficient to ensure accountability. For example, an autonomous aircraft landing system may have fewer safety incidents than human pilots, or a clinical diagnostic support tool may reduce mortality. Statistical evidence might be an appropriate accountability mechanism in many AI systems. This is because it both protects trade secrets and can identify widespread but low-risk harms that only become apparent in aggregate (Barocas and Selbst, 2016[29]; Crawford, 2016[30]). Questions of bias or discrimination can be ascertained statistically: for example, a loan approval system might demonstrate its bias by approving more loans for men than women when other factors are controlled for. The permissible error rate and uncertainty tolerated varies depending on the application. For example, the error rate permissible for a translation tool may not be acceptable for autonomous driving or medical examinations.

### Optimisation transparency is transparency about a system's goals and results

Another approach to the transparency of AI systems proposes to shift the governance focus from a system's means to its ends: that is, moving from requiring the explainability of a system's inner workings to measuring its outcomes – i.e. what the system is "optimised" to do. This would require a declaration of what an AI system is optimised for, with the understanding that optimisations are imperfect, entail trade-offs and should be constrained by "critical constraints", such as safety and fairness. This approach advocates for using AI systems for what they are optimised to do. It invokes existing ethical and legal frameworks, as well as social discussions and political processes where necessary to provide input on what AI systems should be optimised for (Weinberger, 2018[1]).

### Explanation relates to a specific outcome from an AI system

*Explanation* is essential for situations in which fault needs to be determined in a specific instance – a situation that may grow more frequent as AI systems are deployed to make recommendations or decisions currently subject to human discretion (Burgess, 2016[31]). The GDPR mandates that data subjects receive meaningful information about the logic involved, the significance and the envisaged consequences of automated decision-making systems. An explanation does not generally need to provide the full decision-making process of the system. Answering one of the following questions is generally enough (Doshi-Velez et al., 2017[28]):

1. **Main factors in a decision**: For many kinds of decisions, such as custody hearings, qualifying for a loan and pre-trial release, a variety of factors must be considered (or are expressly forbidden from being considered). A list of the factors that were important for an AI prediction – ideally ordered by significance – can help ensure that the right factors were included.

2. **Determinant factors, i.e. factors that decisively affect the outcome**: Sometimes, it is important to know whether a particular factor directed the outcome. Changing a particular factor, such as race in university admissions, can show whether the factor was used correctly.

3. **Why did two similar-looking cases result in different outcomes, or vice versa**? The consistency and integrity of AI-based predictions can be assessed. For example, income should be considered when deciding whether to grant a loan, but it should not be both dispositive and irrelevant in otherwise similar cases.

### Explanation is an active area of research but entails costs and possible trade-offs

Technical research is underway by individual companies, standards bodies, non-profit organisations and public institutions to create AI systems that can explain their predictions. Companies in highly regulated areas such as finance, healthcare and human resources are particularly active to address potential financial, legal and reputational risks lined to predictions made by AI systems. For example, US bank Capital One created a research team in 2016 to find ways of making AI techniques more explainable (Knight, 2017[32]). Companies such as MondoBrain have designed user interfaces to help explain meaningful factors (Box 4.5). Non-profit organisations such as OpenAI are researching approaches to develop explainable AI and audit AI decisions. Publicly funded research is also underway. DARPA, for example, is funding 13 different research groups, working on a range of approaches to making AI more explainable.

---

### Box 4.5. Addressing explainability issues through better-designed user interfaces

Some businesses have started to embed explainability into their solutions so that users better understand the AI processes running in the background. One example is MondoBrain. Based in France, it combines human, collective and artificial intelligence to provide an augmented reality solution for the enterprise. Through the use of interactive data-visualisation dashboards, it evaluates all existing data in a company (from Enterprise Resource Planning, Business Programme Management or Customer Relationship Management software, for instance) and provides prescriptive recommendations based on customers' queries (Figure 4.1). It uses an ML algorithm to eliminate the business variables that provide no value to the query and extract the variables with the most significant impact.

Simple traffic light colours lead users at every step of the query, facilitating their understanding of the decision process. Every single decision is automatically documented, becoming auditable and traceable. It creates a full but simple record of all the steps that led to the final business recommendation.

#### Figure 4.1. Illustration of data-visualisation tools to augment explainability

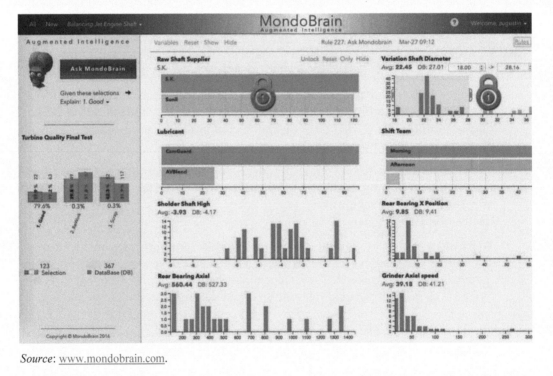

*Source*: www.mondobrain.com.

---

In many cases, it is possible to generate one or more of these kinds of explanations on AI systems' outcomes. However, explanations bear a cost. Designing a system to provide an explanation can be complex and expensive. Requiring explanations for all AI systems may not be appropriate in light of their purpose and may disadvantage SMEs in particular. AI systems must often be designed *ex ante* to provide a certain kind of explanation. Seeking explanations after the fact usually requires additional work, possibly recreating the entire decision system. For example, an AI system cannot provide an explanation of all the top factors that impacted an outcome if it was designed to provide only one. Similarly, an AI system to detect heart conditions cannot be queried about the impact of gender on a diagnosis if gender data were not used to train the system. This is the case even if an AI system actually accounts for gender through proxy variables, such as other medical conditions that are more frequent in women.

In some cases, there is a trade-off between explainability and accuracy. Being explainable may require reducing the solution variables to a set small enough that humans can understand. This could be suboptimal in complex, high-dimensional problems. For example, some ML models used in medical diagnosis can accurately predict the probability of a medical condition, but are too complex for humans to understand. In such cases, the potential harm from a less accurate system that offers clear explanations should be weighed against the potential harm from a more accurate system where errors are harder to detect. For example, recidivism prediction may require simple and explainable models where errors can be detected (Dressel and Farid, 2018[33]). In areas like climate predictions, more complex models that deliver better predictions but are less explainable may be more acceptable. This is particularly the case if other mechanisms to ensure accountability exist, such as statistical data to detect possible bias or error.

## Robustness, security and safety

### *Understanding of robustness, security and safety*

Robustness can be understood as the ability to withstand or overcome adverse conditions (OECD, 2019[34]), including digital security risks. Safe AI systems can be understood as systems that do not pose unreasonable safety risks in normal or foreseeable use or misuse throughout their lifecycle (OECD, 2019[35]). Issues of robustness and safety of AI are interlinked. For example, digital security can affect product safety if connected products such as driverless cars or AI-powered home appliances are not sufficiently secure; hackers could take control of them and change settings at a distance.

### *Risk management in AI systems*

### *Needed level of protections depends on risk-benefit analysis*

The potential harm of an AI system should be balanced against the costs of building transparency and accountability into AI systems. Harms could include risks to human rights, privacy, fairness and robustness. But not every use of AI presents the same risks, and requiring explanation, for example, imposes its own set of costs. In managing risk, there appears to be broad-based agreement that high-stakes' contexts require higher degrees of transparency and accountability, particularly where life and liberty are at stake.

### *Use of risk management approaches throughout the AI lifecycle*

Organisations use risk management to identify, assess, prioritise and treat potential risks that can adversely affect a system's behaviour and outcomes. Such an approach can also be used to identify risks for different stakeholders and determine how to address these risks throughout the AI system lifecycle (Section "AI system lifecycle" in Chapter 1).

AI actors – those who play an active role in the AI system lifecycle – assess and mitigate risks in the AI system as a whole, as well as in each phase of its lifecycle. Risk management in AI systems consists of the following steps, whose relevance varies depending on the phase of the AI system lifecycle:

1. **Objectives**: define objectives, functions or properties of the AI system, in context. These functions and properties may change depending on the phase of the AI lifecycle.

2. **Stakeholders and actors**: identify stakeholders and actors involved, i.e. those directly or indirectly affected by the system's functions or properties in each lifecycle phase.

3. **Risk assessment**: assess the potential effects, both benefits and risks, for stakeholders and actors. These will vary, depending on the stakeholders and actors affected, as well as the phase in the AI system lifecycle.

4. **Risk mitigation**: identify risk mitigation strategies that are appropriate to, and commensurate with, the risk. These should consider factors such as the organisation's goals and objectives, the stakeholders and actors involved, the likelihood of risks manifesting and potential benefits.

5. **Implementation**: implement risk mitigation strategies.

6. **Monitoring, evaluation and feedback**: monitor, evaluate and provide feedback on the results of the implementation.

The use of risk management in AI systems lifecycle and the documentation of the decisions at each lifecycle phase can help improve an AI system's transparency and an organisation's accountability for the system.

### Aggregate harm level should be considered alongside the immediate risk context

Viewed in isolation, some uses of AI systems are low risk. However, they may require higher degrees of robustness because of their societal effects. If a system's operation results in minor harm to a large number of people, it could still collectively cause significant harm overall. Imagine, for example, if a small set of AI tools is embedded across multiple services and sectors. These could be used to obtain loans, qualify for insurance and pass background checks. A single error or bias in one system could create numerous cascading setbacks (Citron and Pasquale, 2014[36]). On its own, any one setback might be minor. Collectively, however, they could be disruptive. This suggests that policy discussions should consider aggregate harm level, in addition to the immediate risk context.

### *Robustness to digital security threats related to AI*

### AI allows more sophisticated attacks of potentially larger magnitude

AI's malicious use is expected to increase as it becomes less expensive and more accessible, in parallel to its uses to improve digital security (Subsection "AI in digital security" in Chapter 3). Cyber attackers are increasing their AI capabilities. Faster and more sophisticated attacks pose a growing concern for digital security.[6] Against this backdrop, existing threats are expanding, new threats are being introduced and the character of threats is changing.

A number of vulnerabilities characterise today's AI systems. Malicious actors can tamper with the data on which an AI system is being trained (e.g. "data poisoning"). They can also identify the characteristics used by a digital security model to flag malware. With this information, they can design unidentifiable malicious code or intentionally cause the misclassification of information (e.g. "adversarial examples") (Box 4.6) (Brundage et al., 2018[37]). As AI technologies are increasingly available, more people can use AI to carry out more sophisticated attacks of potentially larger magnitude. The frequency and efficiency of labour-intensive digital security attacks such as targeted spear phishing could grow as they are automated based on ML algorithms.

**Box 4.6. The peril of adversarial examples for ML**

**Adversarial examples** are inputs fed into ML models that an attacker intentionally designs to cause the model to make a mistake, while displaying a high degree of confidence. Adversarial examples are a real problem for AI robustness and safety because several ML models, including state-of-the-art neural networks, are vulnerable to them.

Adversarial examples can be subtle. In Figure 4.2, an imperceptibly small perturbation or "adversarial input" has been added to the image of a panda. It is specifically designed to trick the image-classification model. Ultimately, the algorithm classifies the panda as a gibbon with close to 100% confidence.

Moreover, recent research has shown that adversarial examples can be created by printing an image on normal paper and photographing it with a standard resolution smart phone. These images could be dangerous: an adversarial sticker on a stop traffic sign could trick a self-driving car into interpreting it as a "yield" or any other sign.

**Figure 4.2. A small perturbation tricks an algorithm into classifying a panda as a gibbon**

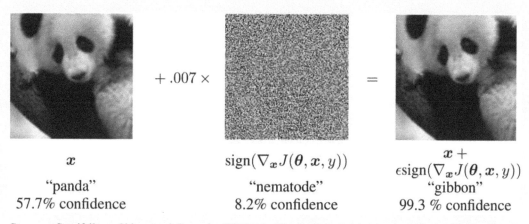

$x$

"panda"
57.7% confidence

$\text{sign}(\nabla_x J(\theta, x, y))$

"nematode"
8.2% confidence

$x + \epsilon\text{sign}(\nabla_x J(\theta, x, y))$

"gibbon"
99.3 % confidence

*Sources*: Goodfellow, Shlens and Szegedy (2015[38]), "Explaining and harnessing adversarial examples", https://arxiv.org/pdf/1412.6572.pdf; Kurakin, Goodfellow and Bengio (2017[39]), "Adversarial examples in the physical world", https://arxiv.org/abs/1607.02533.

*Safety*

### Learning and autonomous AI systems impact policy frameworks for safety

The range of AI-embedded products is expanding rapidly – from robotics and driverless cars to everyday life consumer products and services such as smart appliances and smart home security systems. AI-embedded products offer significant safety benefits, while posing new practical and legal challenges to product safety frameworks (OECD, 2017[20]). Safety frameworks tend to regulate "finished" hardware products rather than software, while a number of AI software products learn and evolve throughout their lifecycle.[7] AI products can also be "autonomous" or semi-autonomous, i.e. make and execute decisions with no or little human input.

Different types of AI applications are expected to call for different policy responses (Freeman, 2017[40]). In broad terms, AI systems call for four considerations. First, they must consider how best to make sure that products are safe. In other words, products must not pose unreasonable safety risk in normal or foreseeable use or misuse throughout their entire lifecycle. This includes cases for which there are few data to train the system on (Box 4.7). Second, they should consider who should be liable, and to what extent, for harm caused by an AI system. At the same time, they should consider which parties can contribute to the safety of autonomous machines. These parties could include users, product and sensor manufacturers, software producers, designers, infrastructure providers and data analytics companies. Third, they should consider the choice of liability principle(s). These could include strict liability, fault-based liability and the role of insurance. The opacity of some AI systems compounds the issue of liability. Fourth, they should consider how the law can be enforced, what is a "defect" in an AI product, what is the burden of proof and what remedies are available.

---

**Box 4.7. Synthetic data for safer and more accurate AI: The case of autonomous vehicles**

The use of synthetic data is gaining widespread adoption in the ML community, as it allows to simulate scenarios that are difficult to observe or replicate in real life. As Philipp Slusallek, Scientific Director at the German Research Centre for Artificial Intelligence explained, one such example is guaranteeing that a self-driving car will not hit a child running across the street.

A "digital reality" – a simulated environment replicating the relevant features of the real world – could have four effects. First, it could generate synthetic input data to train AI systems on complex situations. Second, it could validate performance and recalibrate synthetic data versus real data. Third, it could set up tests, such as for a driver's licence for AVs. Fourth, it could explore the system's decision-making process and the potential outcomes of alternative decisions. To illustrate, this approach has allowed Google to train its self-driving cars with more than 4.8 million simulated kilometres per day (equivalent to more than 500 round trips between New York City and Los Angeles).

*Sources*: Golson (2016[41]), "Google's self-driving cars rack up 3 million simulated miles every day", https://www.theverge.com/2016/2/1/10892020/google-self-driving-simulator-3-million-miles; Slusallek (2018[42]), *Artificial Intelligence and Digital Reality: Do We Need a CERN for AI?*, https://www.oecd-forum.org/channels/722-digitalisation/posts/28452-artificial-intelligence-and-digital-reality-do-we-need-a-cern-for-ai.

---

The European Union's Product Liability Directive (Directive 85/374/EEC) of 1985 establishes the principle of "liability without fault" or "strict liability". According to the principle, if a defective product causes damage to a consumer, the producer is liable even without negligence or fault. The European Commission is reviewing this directive. Preliminary conclusions find the model to be broadly appropriate (Ingels, 2017[43]). However, current and foreseeable AI technologies do impact the concepts of "product", "safety", "defect" and "damage". This makes the burden of proof more difficult.

In the AV sector, the primary concern of policy makers is ensuring safety. Policy work is needed on how to test AVs to ensure they can operate safely. This includes licensing regimes that evaluate the potential of pre-testing AV systems, or requirements that systems monitor the awareness of human drivers in fallback roles. In some cases, licensing is an issue for firms seeking to test vehicles. Governments' openness to testing also varies. There have been some calls for strict liability of manufacturers of AVs. This liability would be based on the controllability of risk. For example, it would recognise that a mere passenger of a driverless car cannot be at fault or have breached a duty of care. Legal experts suggest that even a "registered keeper" concept would not work because the keeper must be able to control the risk (Borges, 2017[44]). Some suggest that insurance could cover the risk of damage by AVs by classifying registered AVs based on risk assessments.

*Working condition safety standards may require updating*

Direct impacts from AI on working conditions may also include the need for new safety protocols. The imperative is growing for new or revised industry standards and technological agreements between management and workers towards reliable, safe and productive workplaces. The European Economic and Social Committee (EESC) recommended for "stakeholders to work together on complementary AI systems and their co-creation in the workplace" (EESC, 2017[45]).

## Accountability

### *AI's growing use calls for accountability for the proper functioning of AI systems*

*Accountability* focuses on being able to place the onus on the appropriate organisations or individuals for the proper functioning of AI systems. Criteria for accountability include respect for principles of human values and fairness, transparency, robustness and safety. Accountability is based on AI actors' individual roles, the context and the state of art. For policy makers, accountability depends on mechanisms that perform several functions. The mechanisms identify the party responsible for a specific recommendation or decision. They correct the recommendation or decision before it is acted on. They could also challenge or appeal the decision after the fact, or even challenge the system responsible for making the decision (Helgason, 1997[46]).

In practice, the accountability of AI systems often hinges on how well a system performs compared to indicators of accuracy or efficiency. Increasingly, measures also include indicators for goals of fairness, safety and robustness. However, such indicators still tend to be less used than measures of efficiency or accuracy. As with all metrics, monitoring and evaluation can be costly. Thus, the types and frequency of measurements must be commensurate with the potential risks and benefits.

### *The required level of accountability depends on the risk context*

Policy approaches depend on context and use case. For example, accountability expectations may be higher for public sector use of AI. This is particularly true in government functions such as security and law enforcement that have the potential for substantial harms. Formal accountability mechanisms are also often required for private-sector applications in transportation, finance and healthcare, which are heavily regulated. In private-sector areas that are less heavily regulated, use of AI is less subject to formal accountability mechanisms. In these cases, technical approaches to transparency and accountability become even more important. They must ensure that systems designed and operated by private-sector actors respect societal norms and legal constraints.

Some applications or decisions may require a human to be "in the loop" to consider the social context and potential unintended consequences. When decisions significantly impact people's lives, there is broad agreement that AI-based outcomes (e.g. a score) should not be the sole decision factor. For example, the GDPR stipulates that a human must be in the loop if a decision has a significant impact on people's lives. For example, humans must be informed if AI is used to sentence criminals, make credit determinations, grant educational opportunities or conduct job screening. In high-stakes' situations, formal accountability mechanisms are often required. For example, a judge using AI for criminal sentencing is a "human-in-the-loop". However, other accountability mechanisms – including a traditional judicial appeals process – help ensure that judges consider AI recommendations as just one element in a prediction (Wachter, Mittelstadt and Floridi, 2017[47]). Low-risk contexts, such as a restaurant recommendation, could rely solely on machines. It may not require such a multi-layered approach, which may impose unnecessary costs.

## AI policy environment

National policies are needed to promote trustworthy AI systems. Such policies can spur beneficial and fair outcomes for people and for the planet, especially in promising areas underserved by market-driven investments. The creation of an enabling policy environment for trustworthy AI includes, among other things, facilitating public and private investment in AI research and development and equipping people with the skills necessary to succeed as jobs evolve. The following subsections explore four policy areas that are critical to the promotion and development of trustworthy AI.

## Investment in AI research and development

### *Long-term investment in public research can help shape AI innovation*

The OECD is considering the role of innovation policies for digital transformation and AI adoption (OECD, 2018[48]). One issue being considered is the role of public research policies, knowledge transfer and co-creation policies to develop research tools and infrastructures for AI. AI calls for policy makers to reconsider the appropriate level of government involvement in AI research to address societal challenges (OECD, 2018[13]). In addition, research institutions in all areas will require capable AI systems to remain competitive, particularly in biomedical science and life science fields. New instruments such as data-sharing platforms and supercomputing facilities can help enable AI research and may call for new investments. Japan, for example, invests more than USD 120 million annually to build a high-performance computing infrastructure for universities and public research centres.

AI is considered to be a general-purpose technology with the potential to impact a large number of industries (Agrawal, Gans and Goldfarb, 2018[49]) (Brynjolfsson, Rock and Syverson, 2017[50]). AI is also considered an "invention of a method of invention" (Cockburn, Henderson and Stern, 2018[51]) that is already widely used by scientists and inventors to facilitate innovation. Entirely new industries could be created based on the scientific breakthroughs enabled by AI. This underscores the importance of basic research and of considering long time horizons in research policy (OECD, 2018[52]).

## Enabling digital ecosystem for AI

### *AI technologies and infrastructure*

Significant advances in AI technologies have taken place over recent years. This has been due to the maturity of statistical modelling techniques such as neural networks, and more particularly, deep neural networks (known as deep learning). Many of the tools to manage and use AI exist as open-source resources in the public domain. This facilitates their adoption and allows for crowdsourcing solutions to software bugs. Tools include TensorFlow (Google), Michelangelo (Uber) and Cognitive Toolkit (Microsoft). Some companies and researchers also share curated training datasets and training tools publicly to help diffuse AI technology.

AI partly owes its recent achievements to the exponential increase in computer speeds and to Moore's Law (i.e. the number of transistors in a dense integrated circuit doubles about every two years). Together, these two developments allow AI algorithms to process enormous amounts of data rapidly. As AI projects move from concept to commercial application, specialised and expensive cloud computing and graphic-processing unit resources are often needed. Trends in AI systems continue to show extraordinary growth in the computational power required. According to one estimate, the largest recent experiment, AlphaGo Zero, required 300 000 times the computing power needed for the largest experiment just six years before (OpenAI, 16 May 2018[53]). AlphaGo Zero's achievements in chess and Go involved computing power estimated to exceed that of the world's ten most powerful supercomputers combined (OECD, 2018[52]).

## *Access to and use of data*

### *Data access and sharing can accelerate or hinder progress in AI*

Current ML technologies require curated and accurate data to train and evolve. Access to high-quality datasets is thus critical to AI development. Factors related to data access and sharing that can accelerate or hinder progress in AI include (OECD, forthcoming[54]):

- **Standards**: Standards are needed to allow interoperability and data re-use across applications, to promote accessibility and to ensure that data are findable, catalogued and/or searchable and re-usable.

- **Risks**: Risks to individuals, organisations and countries of sharing data can include confidentiality and privacy breaches, risks to intellectual property rights (IPRs) and commercial interests, potential national security risks and digital security risks.

- **Costs of data**: Data collection, access, sharing and re-use require up-front and follow-up investments. In addition to data acquisition, additional investments are needed for data cleaning, data curation, metadata maintenance, data storage and processing, and secure IT infrastructure.

- **Incentives**: Market-based approaches can help provide incentives to provide access to, and share data with, data markets and platforms that commercialise data and provide added-value services such as payment and data exchange infrastructure.

- **Uncertainties about data ownership**: Legal frameworks – IPRs, (cyber-) criminal law, competition law and privacy protection law – combined with the involvement of multiple parties in the creation of data have led to uncertainties around the question of "data ownership".

- **User empowerment, including AI-powered agents**: Empowering data users and facilitating data portability – as well as enabling effective consent and choice for data subjects – can encourage individuals and businesses to share personal or business data. Some underscore how AI-powered agents that know individuals' preferences could help them negotiate complex data sharing with other AI systems (Neppel, 2017[55]).

- **Trusted third parties**: Third parties can enable trust and facilitate data sharing and re-use among all stakeholders. Data intermediaries can act as certification authorities. Trusted data-sharing platforms, such as data trusts, provide high-quality data. And institutional review boards assure respect for legitimate interests of third parties.

- **Data representativeness**: AI systems make predictions based on patterns identified in training data sets. In this context, both for accuracy and fairness, training datasets must be inclusive, diverse and representative so they do not under- or misrepresent specific groups.

### *Policies can enhance data access and sharing for the development of AI*

Policy approaches to enhance data access and sharing include (OECD, forthcoming[54]):

- **Providing access to public sector data**, including public sector data, open government data, geo-data (e.g. maps) and transportation data.

- **Facilitating data sharing in the private sector**, usually either on a voluntary basis or, for mandatory policies, restricted data sharing with trusted users. Particular focus areas are "data of public interest", data in network industries such as transportation and energy for service interoperability, and personal data portability.

- **Developing statistical/data analytic capacities**, by establishing technology centres that provide support and guidance in the use and analysis of data.

- **Developing national data strategies**, to ensure the coherence of national data governance frameworks and their compatibility with national AI strategies.

*Technical approaches are emerging to address data constraints*

Some ML algorithms, such as the ones applied to image recognition, exceed average human capabilities. Yet, to get to this point, they had to be trained with large databases of millions of labelled images. The need for data has encouraged active research in machine-learning techniques that require fewer data to train AI systems. Several methods can help address such lack of data.

- **Deep reinforcement learning** is an ML technique that combines deep neural networks with reinforcement learning (Subsection "Cluster 2: ML techniques" in Chapter 1). In this way, it learns to favour a specific behaviour that leads to the desired outcome (Mousave, Schukat and Howley, 2018[56]). Artificially intelligent "agents" compete through actions in a complex environment and receive either a reward or a penalty depending on whether the action led to the desired outcome or not. The agents adjust their actions according to this "feedback".[8]

- **Transfer learning or pre-training** (Pan and Yang, 2010[57]) reuses models that have been trained to perform different tasks in the same domain. For instance, some layers of a model trained to recognise cat pictures could be reused to detect images of blue dresses. In these cases, the sample of images would be orders of magnitude smaller than traditional ML algorithms require (Jain, 2017[58]).

- **Augmented data learning**, or data "synthetisation" can artificially create data through simulations or interpolations based on existing data. This effectively augments these data and improves learning. This method is particularly beneficial in cases where privacy constraints limit data usage or to simulate scenarios seldom encountered in reality (Box 4.7).[9]

- **Hybrid learning models** can model uncertainty by combining different types of deep neural networks with probabilistic or Bayesian approaches. In this way, they can model uncertainty to improve performance and explainability and reduce the likelihood of erroneous predictions (Kendall, 23 May 2017[59]).

Privacy, confidentiality and security concerns may limit data access and sharing. This could lead to a time lag between the speed at which AI systems can learn and the availability of datasets to train them. Recent cryptographic advances, such as in secure multi-party computation (MPC) and homomorphic encryption, could help enable rights-preserving data analyses. Specifically, they could let AI systems operate without collecting or accessing sensitive data (Box 4.8). AI models, in turn, can increasingly work with encrypted data.[10] These solutions are computationally intensive and may thus be difficult to scale (Brundage et al., 2018[37]).

---

**Box 4.8. New cryptographic tools enable privacy-preserving computation**

Advances in encryption have promising application in AI. For example, an ML model could be trained using combined data from multiple organisations. The process would keep data of all participants confidential. This could help overcome barriers related to privacy or confidentiality concerns. The encryption techniques that enable this form of computation – homomorphic encryption and secure MPC – were discovered years and decades ago, respectively. However, they were too inefficient for practical use. Recent algorithmic and implementation advances mean they are increasingly becoming practical tools that can perform productive analyses on real-world datasets.

- **Homomorphic encryption**: obliviously performing computation on encrypted data without needing to view the unencrypted data.

- **Secure MPC**: computing a function of data collected from many sources without revealing information about any source's data to any other source. Secure MPC protocols allow multiple parties to jointly compute algorithms, while keeping each party's input to the algorithm private.

*Sources*: Brundage et al. (2018[37]), *The Malicious Use of Artificial Intelligence: Forecasting, Prevention, and Mitigation*, https://arxiv.org/ftp/arxiv/papers/1802/1802.07228.pdf; Dowlin (2016[60]), *CryptoNets: Applying Neural Networks to Encrypted Data with High Throughput and Accuracy*, https://www.microsoft.com/en-us/research/wp-content/uploads/2016/04/CryptonetsTechReport.pdf.

---

Alternatively, AI models could leverage blockchain technologies that also use cryptographic tools to provide secure data storage (Box 4.9). Solutions combining AI and blockchain technologies could help increase the availability of data. At the same time, they could minimise the privacy and security risks related to unencrypted data processing.

---

**Box 4.9. Blockchain for privacy-preserving identity verification in AI**

Kairos, a face-recognition enterprise solution, has incorporated blockchain technologies into its portfolio. It combines face biometrics and blockchain technology to allow users to better protect their privacy. An algorithm compares a person's image with featured facial landmarks (or identifiers) until a unique match is constructed. This match is then converted into a unique and random string of numbers, after which the original image can be discarded. This "biometric blockchain" is built under the premise that businesses or governments do not need to know who you are to verify that it is, actually, you.

*Source*: https://kairos.com/.

---

## Competition

The OECD has researched the impact and policy implications of the digital transformation on competition (OECD, 2019[61]). This subsection outlines a few possible impacts on competition particularly caused by AI. It recognises the wide recognition for the procompetitive effects of AI in facilitating new entry. It also notes that much of the attention of competition policy given to large AI players is due to their role as online platforms and holders of large amounts of data. It is not connected to their use of AI as such.

A question that relates to AI more specifically is whether there is a data-driven network effect. Under such an effect, each user's utility from using certain kinds of platforms increases whenever others use it, too. By using one of these platforms, for example, users are helping teach its algorithms how to become better at serving users (OECD, 2019[62]). Others have put forward that data exhibit decreasing returns to scale: prediction improvements become more and more marginal as data increase beyond a certain threshold. As a result, some have questioned whether AI could generate long-term competition concerns (Bajari et al., 2018[63]; OECD, 2016[64]; Varian, 2018[65]).

There may be economies of scale in terms of the business value of additional data. If a slight lead in data quality over its competitors enables a company to get many more customers, it could generate a positive feedback loop. More customers mean more data, reinforcing the cycle and allowing for increased market dominance over time. There may also be economies of scale with respect to the expertise required to build effective AI systems.

There is also a concern that algorithms could facilitate collusion through monitoring market conditions, prices and competitors' responses to price changes. These actions could provide companies with new and improved tools for co-ordinating strategies, fixing prices and enforcing cartel agreements. A somewhat more speculative concern is that more sophisticated deep-learning algorithms would not even require actual agreements among competitors to arrive at cartel-like outcomes. Instead, these would be achieved without human intervention. That would present difficult enforcement challenges. Competition laws require evidence of agreements or a "meeting of the minds" before a cartel violation can be established and punished (OECD, 2017[66]).

## *Intellectual property*

This subsection outlines a few possible impacts to intellectual property (IP) caused by AI. It notes this is a rapidly evolving area where evidence-based analytical work is only beginning. IP rules generally accelerate the degree and speed of discovery, invention and diffusion of new technology with regard to AI. In this way, they are similar to rules for other technologies protected by IP rights. While IP rules should reward inventors, authors, artists and brand owners, IP policy should also consider AI's potential as an input for further innovation.

The protection of AI with IPRs other than trade secrets may raise new issues on how to incentivise innovators to disclose AI innovations, including algorithms and their training. A European Parliament Office conference discussed three possible types of AI patenting (EPO, 2018[67]). The first type, Core AI, is often related to algorithms, which as mathematical methods are not patentable. In the second type – trained models/ML – claiming variations and ranges might be an issue. Finally, AI could be patented as a tool in an applied field, defined via technical effects. Other international organisations and OECD countries are also exploring the impact of AI in the IP space.[11]

Another consideration raised by the diffusion of AI is whether IP systems need adjustments in a world in which AI systems can themselves make inventions (OECD, 2017[68]). Certain AI systems can already produce patentable inventions, notably in chemistry, pharmaceuticals and biotechnology. In these fields, many inventions consist of creating original combinations of molecules to form new compounds, or in identifying new properties of existing molecules. For example, KnIT, an ML tool developed by IBM, successfully identified kinases – enzymes that act as a catalyst for the transfer of phosphate groups to specific substrates. These kinases had specific properties among a set of known kinases, which were tested experimentally. Software discovered the specific properties of those molecules, and patents were filed for the inventions. These and other matters regarding AI and IP are being considered by expert agencies of OECD countries such as the European Patent Office and the US Patent and

Trademark Office, as well as by the World Intellectual Property Organization. They could also consider issues of copyright protection of AI-processed data.

### *Small and medium-sized enterprises*

Policies and programmes to help SMEs navigate the AI transition are an increasing priority. This is a rapidly evolving area where evidence-based analytical work is beginning. Potential tools to enable digital ecosystems for SMEs to adopt and leverage AI include:

- Upskilling, which is viewed as critical because competing for scarce AI talent is a particular concern for SMEs.

- Encouraging targeted investments in selected vertical industries. Policies to encourage investment in specific AI applications in French agriculture, for example, could benefit all players where individual SMEs could not afford to invest alone (OECD, 2018[13]).

- Helping SMEs to access data, including by creating platforms for data exchange.

- Supporting SMEs' improved access to AI technologies, including through technology transfer from public research institutes, as well as their access to computing capacities and cloud platforms (Germany, 2018[69]).

- Improving financing mechanisms to help AI SMEs scale up, e.g. through a new public investment fund and increasing the flexibility and financial limits of schemes to invest in knowledge-intensive companies (UK, 2017[70]). The European Commission is also focusing on supporting European SMEs, including through its AI4EU project, an AI-on-demand platform.

### Policy environment for AI innovation

The OECD is analysing changes to innovation and other AI-relevant policies needed in the context of AI and other digital transformations (OECD, 2018[48]). Under consideration is how to improve the adaptability, reactivity and versatility of policy instruments and experiments. Governments can use experimentation to provide controlled environments for the testing of AI systems. Such environments could include regulatory sandboxes, innovation centres and policy labs. Policy experiments can operate in "start-up mode". In this case, experiments are deployed, evaluated and modified, and then scaled up or down, or abandoned quickly.

Another option to spur faster and more effective decisions is the use of digital tools to design policy, including innovation policy, and to monitor policy targets. For instance, some governments use "agent-based modelling" to anticipate the impact of policy variants on different types of businesses.

Governments can encourage AI actors to develop self-regulatory mechanisms such as codes of conduct, voluntary standards and best practices. These can help guide AI actors through the AI lifecycle, including for monitoring, reporting, assessing and addressing harmful effects or misuse of AI systems.

Governments can also establish and encourage public- and private-sector oversight mechanisms of AI systems, as appropriate. These could include compliance reviews, audits, conformity assessments and certification schemes. Such mechanisms could be used while considering the specific needs of SMEs and the constraints they face.

### Jobs

> AI is expected to complement humans in some tasks, replace them in others and generate new types of work

The OECD has researched the impact of the broader digital transformation on jobs and the policy implications in depth (OECD, 2019[61]). As a rapidly evolving area where evidence-based analytical work is beginning, AI is broadly expected to change the nature of work as it diffuses across sectors. It will complement humans in some tasks, replace them in others and also generate new types of work. This section outlines some anticipated changes to labour markets caused by AI, as well as policy considerations to accompany the transition to an AI economy.

### AI is expected to improve productivity

AI is expected to improve productivity in two ways. First, some activities previously carried out by people will be automated. Second, through machine autonomy, systems will operate and adapt to circumstances with reduced or no human control (OECD, 2017[68]; Autor and Salomons, 2018[71]). Research on 12 developed economies estimated that AI could increase labour productivity by up to 40% by 2035 compared to expected baseline levels (Purdy and Daugherty, 2016[72]). Examples abound. IBM's Watson assists client advisors at Crédit Mutuel, a French bank, to field client questions 60% faster.[12] Alibaba's chatbot handled more than 95% of customer inquiries during a 2017 sale. This allowed human customer representatives to handle more complicated or personal issues (Zeng, 2018[73]). In theory, increasing worker productivity should result in higher wages, since each individual employee produces more value added.

Human-AI teams help mitigate error and could expand opportunities for human workers. Human-AI teams have been found to be more productive than either AI or workers alone (Daugherty and Wilson, 2018[74]). For example, human-AI teams in BMW factories increased manufacturing productivity by 85% compared to non-integrated teams. Beyond manufacturing, Walmart robots scan for inventory, leaving store associates to focus on helping customers. And when a human radiologist combined with AI models to screen chest X-rays for tuberculosis, net accuracy reached 100% – higher than AI or human methods alone (Lakhani and Sundaram, 2017[75]).

AI can also make previously automated tasks work better and faster. As a result, companies can produce more at lower cost. If lower costs are passed down to companies or individuals, demand for the goods can be expected to increase. This boosts labour demand both in the company – for instance, in production-related roles – as well as in downstream sectors for intermediate goods.

### AI is expected to change – perhaps accelerate – the tasks that can be automated

Automation is not a new phenomenon, but AI is expected to change, and perhaps accelerate, the profile of tasks that can be automated. Unlike computers, AI technologies are not strictly pre-programmed and rules-based. Computers have tended to reduce employment in routine, middle-skill occupations. However, new AI-powered applications can increasingly perform relatively complex tasks that involve making predictions (see Chapter 3). These tasks include transcription, translation, driving vehicles, diagnosing illness and answering customer inquiries

(Graetz and Michaels, 2018[76]; Michaels, Natraj and Van Reenen, 2014[77]; Goos, Manning and Salomons, 2014[78]).[13]

Exploratory OECD measurement estimated the extent to which technologies can answer the literacy and numeracy questions of the *OECD Survey of Adult Skills* (*PIAAC*) (Elliott, 2017[79]). This research suggested that, in 2017, AI systems could answer the literacy questions at a level comparable to that of 89% of adults in OECD countries. In other words, only 11% of adults were above the level that AI was close to reproducing in terms of literacy skills. The report predicted more economic pressure to apply computer capabilities for certain literacy and numeracy skills. This would likely decrease demand for human workers to perform tasks using low- and mid-level literacy skills, reversing recent patterns. The report underscored the difficulty of designing education policies to adults above the current computer level. It suggested new tools and incentives for promoting adult skills or combining skills policies with other interventions, including social protection and social dialogue (OECD, 2018[13]).

### AI's impact on jobs will depend on its speed of diffusion across different sectors

AI's impact on jobs will also depend on the speed of the development and diffusion of AI technologies in different sectors over the coming decades. AVs are widely expected to disrupt driving and delivery service jobs. Established truck companies such as Volvo and Daimler, for example, are competing with start-ups like Kodiak and Einride to develop and test driverless trucks (Stewart, 2018[80]). According to the International Transport Forum, driverless trucks may be a regular presence on many roads within the next ten years. Some 50% to 70% of the 6.4 million professional trucking jobs in the United States and Europe could be eliminated by 2030 (ITF, 2017[81]). However, new jobs will be created in parallel to provide support services for the increased number of driverless trucks. Driverless trucks could reduce operating costs for road freight in the order of 30%, notably due to savings in labour costs. This could drive traditional trucking companies out of business, resulting in an even faster decline in trucking jobs.

### AI technologies are likely to impact traditionally higher-skilled tasks

AI technologies are also performing prediction tasks traditionally performed by higher-skilled workers, from lawyers to medical personnel. A robolawyer has successfully appealed over USD 12 million worth of traffic tickets (Dormehl, 2018[82]). In 2016, IBM's Watson and DeepMind Health outperformed human doctors in diagnosing rare cancers (Frey and Osborne, 2017[83]). AI has proven to be better at predicting stock exchange variations than finance professionals (Mims, 2010[84]).

### AI can complement people and create new types of work

AI complements people and is also likely to create job opportunities for human workers. Notable areas include those that complement prediction and leverage human skills such as critical thinking, creativity and empathy (EOP, 2016[85]; OECD, 2017[20]).

- **Data scientists and ML experts**: Specialists are needed to create and clean data and to program and develop AI applications. However, although data and ML lead to some new tasks, they are unlikely to generate large numbers of new tasks for workers.

- **Actions**: Some actions are inherently more valuable when done by a human than a machine, as professional athletes, child carers or salespeople illustrate. Many think it is likely that humans will increasingly focus on work to improve each other's lives, such as childcare, physical coaching and care for the terminally ill.

- **Judgment to determine what to predict**: Perhaps most important is the concept of judgment – the process of determining the reward to a particular action in a particular environment. When AI is used for predictions, a human must decide what to predict and what to do with the predictions. Posing dilemmas, interpreting situations or extracting meaning from text requires people with qualities such as judgment and fairness (OECD, 2018[13]). In science, for example, AI can complement humans in charge of the conceptual thinking necessary to build research frameworks and to set the context for specific experiments.

- **Judgment to decide what to do with a prediction**: A decision cannot be made with a prediction alone. For example, the trivial decision of whether to take an umbrella when going outside for a walk will consider a prediction about the likelihood of rain. However, the decision will depend largely on preferences such as the degree to which one dislikes being wet and carrying an umbrella. This example can be broadened to many important decisions. In cybersecurity, a prediction about whether a new inquiry is hostile will need to be measured against the risk of turning away a friendly inquiry and letting a hostile inquiry obtain unauthorised information.

### Predictions regarding AI's net impact on the quantity of work vary widely

Over the past five years, widely varying estimates have been made of the overall impacts of automation on job loss (Winick, 2018[86]; MGI, 2017[87]; Frey and Osborne, 2017[83]). For example, a Frey and Osborne predicted that 47% of US jobs are at risk of displacement in the next 10 to 15 years. Using a task-oriented approach, the McKinsey Global Institute found in 2017 that about one-third of activities in 60% of jobs are automatable. However, identified jobs affected by automation are not due to the development and deployment of AI alone, but also to other technological developments.

In addition, anticipating future job creation in new areas is challenging. One study estimated that AI would lead to a net job creation of 2 million by 2025 (Gartner, 2017[88]). Job creation is likely both as a result of new occupations arising and through more indirect channels. For example, AI is likely to reduce the cost of producing goods and services, as well as to increase their quality. This will lead to increased demand and, as a result, higher employment.

The most recent OECD estimates allow for heterogeneity of tasks within narrowly defined occupations, using data of the Programme for the International Assessment of Adult Competencies (PIAAC). Based on existing technologies, 14% of jobs in member countries are at high risk of automation; another 32% of workers are likely to see substantial change in how their jobs are carried out (Nedelkoska and Quintini, 2018[89]). The risk of automation is highest among teenagers and senior workers. Recent OECD analysis finds employment decline in occupations classified as "highly automatable" in 82% of regions across 16 European countries. At the same time, it identifies a greater increase in "low automation" jobs in 60% of regions that offsets job loss. This research supports the idea that automation may be shifting the mix of jobs, without driving down overall employment (OECD, 2018[90]).

### AI will change the nature of work

AI adoption is broadly expected to change the nature of work. AI may help make work more interesting by automating routine tasks, allowing more flexible work and possibly a better work-life balance. Human creativity and ingenuity can leverage increasingly powerful computation, data and algorithm resources to create new tasks and directions that require human creativity (Kasparov, 2018[91]).

More broadly, AI may accelerate changes to how the labour market operates by increasing efficiency. Today, AI techniques coupled with big data hold potential to help companies to identify roles for workers – as well as participate in matching people to jobs. IBM, for example, uses AI to optimise employee training, recommending training modules to employees based on their past performance, career goals and IBM skills needs. Companies such as KeenCorp and Vibe have developed text analytics techniques to help companies parse employee communications to help assess metrics such as morale, worker productivity and network effects (Deloitte, 2017[92]). As a result of this information, AI may help companies optimise worker productivity.

*Parameters for organisational change will need to be set*

The imperative is growing for new or revised industry standards and technological agreements between management and workers towards reliable, safe and productive workplaces. The EESC recommended for "stakeholders to work together on complementary AI systems and their co-creation in the workplace" (EESC, 2017[45]). Workplaces also need flexibility, while safeguarding workers' autonomy and job quality, including the sharing of profits. The recent collective agreement between the German sector union *IG Metall* and employers (*Gesamtmetall)* gives an economic case for variable working times. It shows that, depending on organisational and personal (care) needs in the new world of work, employers and unions can reach agreements without revising legal employment protections (Byhovskaya, 2018[93]).

*Using AI to support labour market functions – with safeguards – is also promising*

AI has already begun to make job matching and training more efficient. It can help better connect job seekers, including displaced workers, with the workforce development programmes they need to qualify for emerging and expanding occupations. In many OECD countries, employers and public employment services already use online platforms to fill jobs (OECD, 2018[90]). Looking ahead, AI and other digital technologies can improve innovative and personalised approaches to job-search and hiring processes and enhance the efficiency of labour supply and demand matching. The LinkedIn platform uses AI to help recruiters find the right candidates and to connect candidates to the right jobs. It draws on data about the profile and activity of the platform's 470 million registered users (Wong, 2017[94]).

AI technologies leveraging big data can also help inform governments, employers and workers about local labour market conditions. This information can help identify and forecast skills demands, direct training resources and connect individuals with jobs. Projects to develop labour market information are already underway in countries such as Finland, the Czech Republic and Latvia (OECD, 2018[90]).

*Governing the use of workers' data*

While AI requires large datasets to be productive, there are some potential risks when these data represent individual workers, especially if the AI systems that analyse the data are opaque. Human resources and productivity planning will increasingly leverage employee data and algorithms. As they do, public policy makers and stakeholders could investigate how data collection and processing affect employment prospects and terms. Data may be collected from applications, fingerprints, wearables and sensors in real time, indicating the location and workplace of an employee. In customer service, AI software analyses the friendliness of employees' tone. According to workers' accounts, however, it did not consider speech patterns and challenging the scoring was difficult (UNI, 2018[95]).

In contrast, agreements on workers' data and the right to disconnect are emerging in some countries. The French telecommunications company Orange France Telecom and five trade union centres were among the first to settle on commitments to protect employee data. Specific protections include transparency over use, training and the introduction of new equipment. To close the regulatory gap on workers' data, provisions could include establishing data governance bodies in companies, accountability on behalf of (personal) data use, data portability, explanation and deletion rights (UNI, 2018[95]).

### Managing the AI transition

#### Policies for managing the AI transition, including social protection, are key

There is a possibility of disruption and turbulence in labour markets as technology outpaces organisational adaptation (OECD, 2018[13]). Long-term optimism does not imply a smooth transition to an economy with more and more AI: some sectors are likely to grow, while others decline. Existing jobs may disappear, while new ones are created. Thus, key policy questions with respect to AI and jobs relate to managing the transition. Policies for managing the transition include social safety nets, health insurance, progressive taxation of labour and capital, and education. Moreover, OECD analysis also points to the need for attention to competition policies and other policies that might affect concentration, market power and income distribution (OECD, 2019[61]).

### Skills to use AI

#### As jobs change, so will the skills required of workers

As jobs change, so will the skills required of workers (OECD, 2017[96]; Acemoglu and Restrepo, 2018[97]; Brynjolfsson and Mitchell, 2017[98]). The present subsection outlines a few possible repercussions of AI on skills, noting this is a rapidly evolving area where evidence-based analytical work is only beginning. Education policy is expected to require adjustments to expand lifelong learning, training and skills development. As with other areas of technology, AI is expected to generate demand in three skills areas. First, **specialist skills** will be needed to program and develop AI applications. These could include skills for AI-related fundamental research, engineering and applications, as well as data science and computational thinking. Second, **generic skills** will be needed to leverage AI, including through AI-human teams on the factory floor and quality control. Third, AI will need **complementarity skills**. These could include leveraging human skills such as critical thinking; creativity, innovation and entrepreneurship; and empathy (EOP, 2016[85]; OECD, 2017[20]).

#### Initiatives to build and develop AI skills are required to address AI skills shortage

The AI skills shortage is expected to grow, and may become more evident as demand for specialists in areas such as ML accelerates. SMEs, public universities and research centres already compete with dominant firms for talent. Initiatives to build and develop AI skills are starting to emerge in the public, private and academic sectors. For instance, the Singaporean government has set up a five-year research programme on governance of AI and data use in Singapore Management University. Its Centre for AI & Data Governance focuses on industry-relevant research, covering AI and industry, society and commercialisation. On the academic side, the Massachusetts Institute of Technology (MIT) has committed USD 1 billion to create the Schwarzman College of Computing. It aims to equip students and researchers in all disciplines to use computing and AI to advance their disciplines and vice versa.

The AI skills shortage has also led some countries to streamline immigration processes for high-skilled experts. For example, the United Kingdom doubled the number of its Tier 1 (Exceptional Talent) visas to 2 000 a year and streamlined the process for top students and researchers to work there (UK, 2017[99]). Similarly, Canada introduced two-week processing times for visa applications from high-skilled workers and visa exemptions for short-term research assignments. This was part of its 2017 Global Skills Strategy to attract high-skilled workers and researchers from abroad (Canada, 2017[100]).

### Generic skills to be able to leverage AI

All OECD countries assess skills and anticipate need for skills in the current, medium or long term. Finland proposed the Artificial Intelligence Programme, which includes a skills account or voucher-based lifelong learning programme to create demand for education and training (Finland, 2017[101]). The United Kingdom is promoting a diverse AI workforce and investing about GBP 406 million (USD 530 million) in skills. It focuses on science, technology, engineering and mathematics, and computer science teachers (UK, 2017[99]).

Practitioners must now be what some call "bilinguals". These are people who may be specialised in one area such as economics, biology or law, but who are also skilled at AI techniques such as ML. In this vein, the MIT announced in October 2018 the most significant change to its structure in 50 years. It plans a new school of computing that will sit outside the engineering discipline and intertwine with all other academic departments. It will train these "bilingual" students who apply AI and ML to the challenges of their own disciplines. This represents a complete shift in the way the MIT teaches computer science. The MIT is allocating USD 1 billion for the creation of this new college within the Institute (MIT, 2018[102]).

### Complementary skills

There is a strong focus on emerging, "softer" skills. Based on existing research, these skills may include human judgment, analysis and interpersonal communication (Agrawal, Gans and Goldfarb, 2018[103]; Deming, 2017[104]; Trajtenberg, 2018[105]). In 2021, the OECD will include a module on the Programme for International Student Assessment (PISA) to test creative and critical thinking skills. The results will help provide a benchmark creativity assessment across countries to inform policy and social partner actions.

## Measurement

The implementation of human-centred and trustworthy AI depends on context. However, a key part of policy makers' commitment to ensuring human-centred AI will be to identify objectives and metrics to assess performance of AI systems. These include areas such as accuracy, efficiency, advancement of societal goals, fairness and robustness.

## References

Abrams, M. et al. (2017), *Artificial Intelligence, Ethics and Enhanced Data Stewardship*, The Information Accountability Foundation, Plano, Texas. [16]

Acemoglu, D. and P. Restrepo (2018), *Artificial Intelligence, Automation and Work*, National Bureau of Economic Research , Cambridge, MA, http://dx.doi.org/10.3386/w24196. [97]

Agrawal, A., J. Gans and A. Goldfarb (2018), "Economic policy for artificial intelligence", *NBER Working Paper*, No. 24690, http://dx.doi.org/10.3386/w24690. [49]

Agrawal, A., J. Gans and A. Goldfarb (2018), *Prediction Machines: The Simple Economics of Artificial Intelligence*, Harvard Business School Press, Brighton, MA. [103]

Autor, D. and A. Salomons (2018), "Is automation labor-displacing? Productivity growth, employment, and the labor share", *NBER Working Paper*, No. 24871, http://dx.doi.org/10.3386/w24871. [71]

Bajari, P. et al. (2018), "The impact of big data on firm performance: An empirical investigation", *NBER Working Paper*, No. 24334, http://dx.doi.org/10.3386/w24334. [63]

Barocas, S. and A. Selbst (2016), "Big data's disparate impact", *California Law Review*, Vol. 104, pp. 671-729, http://www.californialawreview.org/wp-content/uploads/2016/06/2Barocas-Selbst.pdf. [29]

Berk, R. and J. Hyatt (2015), "Machine learning forecasts of risk to inform sentencing decisions", *Federal Sentencing Reporter*, Vol. 27/4, pp. 222-228, http://dx.doi.org/10.1525/fsr.2015.27.4.222. [23]

Borges, G. (2017), *Liability for Machine-Made Decisions: Gaps and Potential Solutions*, presentation at the "AI: Intelligent Machines, Smart Policies" conference, Paris, 26-27 October, http://www.oecd.org/going-digital/ai-intelligent-machines-smart-policies/conference-agenda/ai-intelligent-machines-smart-policies-borges.pdf. [44]

Brundage, M. et al. (2018), *The Malicious Use of Artificial Intelligence: Forecasting, Prevention, and Mitigation*, Future of Humanity Institute, University of Oxford, Centre for the Study of Existential Risk, University of Cambridge, Centre for a New American Security, Electronic Frontier Foundation and Open AI, https://arxiv.org/ftp/arxiv/papers/1802/1802.07228.pdf. [37]

Brynjolfsson, E. and T. Mitchell (2017), "What can machine learning do? Workforce implications", *Science*, Vol. 358/6370, pp. 1530-1534, http://dx.doi.org/10.1126/science.aap8062. [98]

Brynjolfsson, E., D. Rock and C. Syverson (2017), "Artificial intelligence and the modern productivity paradox: A clash of expectations and statistics", *NBER Working Paper*, No. 24001, http://dx.doi.org/10.3386/w24001. [50]

Burgess, M. (2016), "Holding AI to account: Will algorithms ever be free of bias if they are created by humans?", *WIRED*, 11 January, https://www.wired.co.uk/article/creating-transparent-ai-algorithms-machine-learning. [31]

Byhovskaya, A. (2018), *Overview of the National Strategies on Work 4.0: A Coherent Analysis of the Role of the Social Partners*, European Economic and Social Committee, Brussels, https://www.eesc.europa.eu/sites/default/files/files/qe-02-18-923-en-n.pdf. [93]

Canada (2017), "Government of Canada launches the Global Skills Strategy", News Release, Immigration, Refugees and Citizenship Canada, 12 June, https://www.canada.ca/en/immigration-refugees-citizenship/news/2017/06/government_of_canadalaunchestheglobalskillsstrategy.html. [100]

Cellarius, M. (2017), *Artificial Intelligence and the Right to Informational Self-determination*, The OECD Forum, OECD, Paris, https://www.oecd-forum.org/users/75927-mathias-cellarius/posts/28608-artificial-intelligence-and-the-right-to-informational-self-determination. [9]

Chouldechova, A. (2016), "Fair prediction with disparate impact: A study of bias in recidivism prediction instruments", *arXiv*, Vol. 07524, https://arxiv.org/abs/1610.07524. [24]

Citron, D. and F. Pasquale (2014), "The scored society: Due process for automated predictions", *Washington Law Review*, Vol. 89, https://papers.ssrn.com/sol3/papers.cfm?abstract_id=2376209. [36]

Cockburn, I., R. Henderson and S. Stern (2018), "The impact of artificial intelligence on innovation", *NBER Working Paper*, No. 24449, http://dx.doi.org/10.3386/w24449. [51]

Crawford, K. (2016), "Artificial intelligence's white guy problem", *The New York Times*, 26 June, https://www.nytimes.com/2016/06/26/opinion/sunday/artificial-intelligences-white-guy-problem.html?_r=0. [30]

Daugherty, P. and H. Wilson (2018), *Human Machine: Reimagining Work in the Age of AI*, Harvard Business Review Press, Cambridge, MA. [74]

Deloitte (2017), *HR Technology Disruptions for 2018: Productivity, Design and Intelligence Reign*, Deloitte, http://marketing.bersin.com/rs/976-LMP-699/images/HRTechDisruptions2018-Report-100517.pdf. [92]

Deming, D. (2017), "The growing importance of social skills in the labor market", *The Quarterly Journal of Economics*, Vol. 132/4, pp. 1593-1640, http://dx.doi.org/10.1093/qje/qjx022. [104]

Dormehl, L. (2018), "Meet the British whiz kid who fights for justice with robo-lawyer sidekick", *Digital Trends*, 3 March, https://www.digitaltrends.com/cool-tech/robot-lawyer-free-acess-justice/. [82]

Doshi-Velez, F. et al. (2017), "Accountability of AI under the law: The role of explanation", *arXiv* 21 November, https://arxiv.org/pdf/1711.01134.pdf. [28]

Dowlin, N. (2016), *CryptoNets: Applying Neural Networks to Encrypted Data with High Throughput and Accuracy*, Microsoft Research, https://www.microsoft.com/en-us/research/wp-content/uploads/2016/04/CryptonetsTechReport.pdf. [60]

Dressel, J. and H. Farid (2018), "The accuracy, fairness and limits of predicting recidivism", *Science Advances*, Vol. 4/1, http://advances.sciencemag.org/content/4/1/eaao5580. [33]

EESC (2017), *Artificial Intelligence – The Consequences of Artificial Intelligence on the (Digital) Single Market, Production, Consumption, Employment and Society*, European Economic and Social Committee, Brussels, https://www.eesc.europa.eu/en/our-work/opinions-inf. [45]

Elliott, S. (2017), *Computers and the Future of Skill Demand*, Educational Research and Innovation, OECD Publishing, Paris, http://dx.doi.org/10.1787/9789264284395-en. [79]

EOP (2016), *Artificial Intelligence, Automation and the Economy*, Executive Office of the President, Government of the United States, https://www.whitehouse.gov/sites/whitehouse.gov/files/images/EMBARGOED AI Economy Report.pdf. [85]

EPO (2018), *Patenting Artificial Intelligence - Conference Summary*, European Patent Office, Munich, 30 May, http://documents.epo.org/projects/babylon/acad.nsf/0/D9F20464038C0753C125829E0031B8 14/$FILE/summary_conference_artificial_intelligence_en.pdf. [67]

Finland (2017), *Finland's Age of Artificial Intelligence - Turning Finland into a Leader in the Application of AI*, webpage, Finnish Ministry of Economic Affairs and Employment, https://tem.fi/en/artificial-intelligence-programme. [101]

Flanagan, M., D. Howe and H. Nissenbaum (2008), "Embodying values in technology: Theory and practice", in van den Hoven, J. and J. Weckert (eds.), *Information Technology and Moral Philosophy*, Cambridge University Press, Cambridge, http://dx.doi.org/10.1017/cbo9780511498725.017. [15]

Freeman, R. (2017), *Evolution or Revolution? The Future of Regulation and Liability for AI*, presentation at the "AI: Intelligent Machines, Smart Policies" conference, Paris, 26-27 October, http://www.oecd.org/going-digital/ai-intelligent-machines-smart-policies/conference-agenda/ai-intelligent-machines-smart-policies-freeman.pdf. [40]

Frey, C. and M. Osborne (2017), "The future of employment: How susceptible are Jobs to computerisation?", *Technological Forecasting and Social Change*, Vol. 114, pp. 254-280, http://dx.doi.org/10.1016/j.techfore.2016.08.019. [83]

Gartner (2017), "Gartner says by 2020, artificial intelligence will create more jobs than it eliminates", Gartner, Press Release, 13 December, https://www.gartner.com/en/newsroom/press-releases/2017-12-13-gartner-says-by-2020-artificial-intelligence-will-create-more-jobs-than-it-eliminates. [88]

Germany (2018), "Key points for a federal government strategy on artificial intelligence", Press Release, 18 July, BMWI, https://www.bmwi.de/Redaktion/EN/Pressemitteilungen/2018/20180718-key-points-for-federal-government-strategy-on-artificial-intelligence.html. [69]

Golson, J. (2016), "Google's self-driving cars rack up 3 million simulated miles every day", *The Verge*, 1 February, https://www.theverge.com/2016/2/1/10892020/google-self-driving-simulator-3-million-miles. [41]

Goodfellow, I., J. Shlens and C. Szegedy (2015), "Explaining and harnessing adversarial examples", *arXiv*, Vol. 1412.6572, https://arxiv.org/pdf/1412.6572.pdf. [38]

Goos, M., A. Manning and A. Salomons (2014), "Explaining job polarization: Routine-biased technological change and offshoring", *American Economic Review*, Vol. 104/8, pp. 2509-2526, http://dx.doi.org/10.1257/aer.104.8.2509. [78]

Graetz, G. and G. Michaels (2018), "Robots at work", *Review of Economics and Statistics*, Vol. 100/5, pp. 753-768, http://dx.doi.org/10.1162/rest_a_00754. [76]

Harkous, H. (2018), "Polisis: Automated analysis and presentation of privacy policies using deep learning", *arXiv* 29 June, https://arxiv.org/pdf/1802.02561.pdf. [14]

Heiner, D. and C. Nguyen (2018), "Amplify Human Ingenuity with Intelligent Technology", *Shaping Human-Centered Artificial Intelligence, A.Ideas Series*, The Forum Network, OECD, Paris, https://www.oecd-forum.org/users/86008-david-heiner-and-carolyn-nguyen/posts/30653-shaping-human-centered-artificial-intelligence. [6]

Helgason, S. (1997), *Towards Performance-Based Accountability: Issues for Discussion*, Public Management Service, OECD Publishing, Paris, http://www.oecd.org/governance/budgeting/1902720.pdf. [46]

Ingels, H. (2017), *Artificial Intelligence and EU Product Liability Law*, presentation at the "AI: Intelligent Machines, Smart Policies" conference, Paris, 26-27 October, http://www.oecd.org/going-digital/ai-intelligent-machines-smart-policies/conference-agenda/ai-intelligent-machines-smart-policies-ingels.pdf. [43]

ITF (2017), "Driverless trucks: New report maps out global action on driver jobs and legal issues", International Transport Forum, https://www.itf-oecd.org/driverless-trucks-new-report-maps-out-global-action-driver-jobs-and-legal-issues. [81]

Jain, S. (2017), "NanoNets : How to use deep learning when you have limited data, Part 2 : Building object detection models with almost no hardware", *Medium,* 30 January, https://medium.com/nanonets/nanonets-how-to-use-deep-learning-when-you-have-limited-data-f68c0b512cab. [58]

Kasparov, G. (2018), *Deep Thinking: Where Machine Intelligence Ends and Human Creativity Begins*, Public Affairs, New York. [91]

Kendall, A. (23 May 2017), "Deep learning is not good enough, we need Bayesian deep learning for safe AI", Alex Kendall blog, https://alexgkendall.com/computer_vision/bayesian_deep_learning_for_safe_ai/. [59]

Knight, W. (2017), "The financial world wants to open AI's black boxes", *MIT Technology Review,* 13 April, https://www.technologyreview.com/s/604122/the-financial-world-wants-to-open-ais-black-boxes/. [32]

Kosack, S. and A. Fung (2014), "Does transparency improve governance?", *Annual Review of Political Science*, Vol. 17, pp. 65-87, https://www.annualreviews.org/doi/pdf/10.1146/annurev-polisci-032210-144356. [26]

Kosinski, M., D. Stillwell and T. Graepel (2013), "Private traits and attributes are predictable from digital records of human behavior", *PNAS,* 11 March, http://www.pnas.org/content/pnas/early/2013/03/06/1218772110.full.pdf. [2]

Kurakin, A., I. Goodfellow and S. Bengio (2017), "Adversarial examples in the physical world", *arXiv* 02533, https://arxiv.org/abs/1607.02533. [39]

Lakhani, P. and B. Sundaram (2017), "Deep learning at chest radiography: Automated classification of pulmonary tuberculosis by using convolutional neural networks", *Radiology*, Vol. 284/2, pp. 574-582, http://dx.doi.org/10.1148/radiol.2017162326. [75]

Matheson, R. (2018), *Artificial intelligence model "learns" from patient data to make cancer treatment less toxic*, 9 August, http://news.mit.edu/2018/artificial-intelligence-model-learns-patient-data-cancer-treatment-less-toxic-0810. [107]

MGI (2017), *Jobs Lost, Jobs Gained: Workforce Transitions in a Time of Automation*, McKinsey Global Institute, New York. [87]

Michaels, G., A. Natraj and J. Van Reenen (2014), "Has ICT polarized skill demand? Evidence [77] from eleven countries over twenty-five years", *Review of Economics and Statistics*, Vol. 96/1, pp. 60-77, http://dx.doi.org/10.1162/rest_a_00366.

Mims, C. (2010), "AI that picks stocks better than the pros", *MIT Technology Review,* 10 June, [84] https://www.technologyreview.com/s/419341/ai-that-picks-stocks-better-than-the-pros/.

MIT (2018), "Cybersecurity's insidious new threat: Workforce stress", *MIT Technology Review,* [102] 7 August, https://www.technologyreview.com/s/611727/cybersecuritys-insidious-new-threat-workforce-stress/.

Mousave, S., M. Schukat and E. Howley (2018), "Deep reinforcement learning: An overview", [56] *arXiv* 1806.08894, https://arxiv.org/abs/1806.08894.

Narayanan, A. (2018), "Tutorial: 21 fairness definitions and their politics", [17] https://www.youtube.com/watch?v=jIXIuYdnyyk.

Nedelkoska, L. and G. Quintini (2018), "Automation, skills use and training", *OECD Social,* [89] *Employment and Migration Working Papers*, No. 202, OECD Publishing, Paris, https://dx.doi.org/10.1787/2e2f4eea-en.

Neppel, C. (2017), *AI: Intelligent Machines, Smart Policies*, presentation at the "AI: Intelligent [55] Machines, Smart Policies" conference, Paris, 26-27 October, http://oe.cd/ai2017.

NITI (2018), *National Strategy for Artificial Intelligence #AIforall*, NITI Aayog, June, [5] http://niti.gov.in/writereaddata/files/document_publication/NationalStrategy-for-AI-Discussion-Paper.pdf.

OECD (2019), *An Introduction to Online Platforms and Their Role in the Digital* [62] *Transformation*, OECD Publishing, Paris, https://dx.doi.org/10.1787/53e5f593-en.

OECD (2019), *Going Digital: Shaping Policies, Improving Lives*, OECD Publishing, Paris, [61] https://dx.doi.org/10.1787/9789264312012-en.

OECD (2019), *Recommendation of the Council on Artificial Intelligence*, OECD, Paris. [35]

OECD (2019), *Scoping Principles to Foster Trust in and Adoption of AI – Proposal by the* [34] *Expert Group on Artificial Intelligence at the OECD (AIGO)*, OECD, Paris, http://oe.cd/ai.

OECD (2018), "AI: Intelligent machines, smart policies: Conference summary", *OECD Digital* [13] *Economy Papers*, No. 270, OECD Publishing, Paris, http://dx.doi.org/10.1787/f1a650d9-en.

OECD (2018), *Job Creation and Local Economic Development 2018: Preparing for the Future* [90] *of Work*, OECD Publishing, Paris, https://dx.doi.org/10.1787/9789264305342-en.

OECD (2018), *OECD Science, Technology and Innovation Outlook 2018: Adapting to* [52] *Technological and Societal Disruption*, OECD Publishing, Paris, https://dx.doi.org/10.1787/sti_in_outlook-2018-en.

OECD (2018), "Perspectives on innovation policies in the digital age", in *OECD Science,* [48] *Technology and Innovation Outlook 2018: Adapting to Technological and Societal Disruption*, OECD Publishing, Paris, https://dx.doi.org/10.1787/sti_in_outlook-2018-8-en.

OECD (2017), *Algorithms and Collusion: Competition Policy in the Digital Age*, OECD [66] Publishing, Paris, http://www.oecd.org/competition/algorithms-collusion-competition-policy-in-the-digital-age.html.

OECD (2017), *Getting Skills Right: Skills for Jobs Indicators*, OECD Publishing, Paris, [96] https://dx.doi.org/10.1787/9789264277878-en.

OECD (2017), *OECD Digital Economy Outlook 2017*, OECD Publishing, Paris, http://dx.doi.org/10.1787/9789264276284-en. [20]

OECD (2017), *The Next Production Revolution: Implications for Governments and Business*, OECD Publishing, Paris, https://dx.doi.org/10.1787/9789264271036-en. [68]

OECD (2016), *Big Data: Bringing Competition Policy to the Digital Era (Executive Summary)*, OECD DAF Competition Committee, https://one.oecd.org/document/DAF/COMP/M(2016)2/ANN4/FINAL/en/pdf. [64]

OECD (2013), *Recommendation of the Council concerning Guidelines Governing the Protection of Privacy and Transborder Flows of Personal Data*, OECD, Paris, http://www.oecd.org/sti/ieconomy/2013-oecd-privacy-guidelines.pdf. [12]

OECD (2011), *OECD Guidelines for Multinational Enterprises, 2011 Edition*, OECD Publishing, Paris, https://dx.doi.org/10.1787/9789264115415-en. [8]

OECD (forthcoming), *Enhanced Access to and Sharing of Data: Reconciling Risks and Benefits for Data Re-Use across Societies*, OECD Publishing, Paris. [54]

OHCHR (2011), *Guiding Principles on Business and Human Rights*, United Nations Human Rights Office of the High Commissioner, https://www.ohchr.org/Documents/Publications/GuidingPrinciplesBusinessHR_EN.pdf. [7]

O'Neil, C. (2016), *Weapons of Math Destruction: How Big Data Increases Inequality and Threatens Democracy*, Broadway Books, New York. [25]

OpenAI (16 May 2018), "AI and compute", OpenAI blog, San Francisco, https://blog.openai.com/ai-and-compute/. [53]

Pan, S. and Q. Yang (2010), "A survey on transfer learning", *IEEE Transactions on Knowledge and Data Engineering*, Vol. 22/10, pp. 1345-1359. [57]

Paper, I. (ed.) (2018), "Artificial intelligence and privacy", June, Office of the Victorian Information Commissioner, https://ovic.vic.gov.au/wp-content/uploads/2018/08/AI-Issues-Paper-V1.1.pdf. [11]

Patki, N., R. Wedge and K. Veeramachaneni (2016), "The Synthetic Data Vault", *2016 IEEE International Conference on Data Science and Advanced Analytics (DSAA)*, http://dx.doi.org/10.1109/dsaa.2016.49. [106]

Privacy International and Article 19 (2018), *Privacy and Freedom of Expression in the Age of Artificial Intelligence*, https://www.article19.org/wp-content/uploads/2018/04/Privacy-and-Freedom-of-Expression-In-the-Age-of-Artificial-Intelligence-1.pdf. [10]

Purdy, M. and P. Daugherty (2016), "Artificial intelligence poised to double annual economic growth rate in 12 developed economies and boost labor productivity by up to 40 percent by 2035, according to new research by Accenture", Accenture, Press Release, 28 September, http://www.accenture.com/futureofAI. [72]

Selbst, A. (2017), "Disparate impact in big data policing", *Georgia Law Review*, Vol. 52/109, http://dx.doi.org/10.2139/ssrn.2819182. [21]

Simonite, T. (2018), "Probing the dark side of Google's ad-targeting system", *MIT Technology Review*, 6 July, https://www.technologyreview.com/s/539021/probing-the-dark-side-of-googles-ad-targeting-system/. [19]

Slusallek, P. (2018), *Artificial Intelligence and Digital Reality: Do We Need a CERN for AI?*, The Forum Network, OECD, Paris, https://www.oecd-forum.org/channels/722-digitalisation/posts/28452-artificial-intelligence-and-digital-reality-do-we-need-a-cern-for-ai. [42]

Smith, M. and S. Neupane (2018), *Artificial Intelligence and Human Development: Toward a Research Agenda*, International Development Research Centre, Ottawa, https://idl-bnc-idrc.dspacedirect.org/handle/10625/56949. [4]

Stewart, J. (2018), "As Uber gives up on self-driving trucks, another startup jumps in", *WIRED*, 8 July, https://www.wired.com/story/kodiak-self-driving-semi-trucks/. [80]

Talbot, D. et al. (2017), "Charting a roadmap to ensure AI benefits all", *Medium*, 30 November, https://medium.com/berkman-klein-center/charting-a-roadmap-to-ensure-artificial-intelligence-ai-benefits-all-e322f23f8b59. [3]

Trajtenberg, M. (2018), "AI as the next GPT: A political-economy perspective", *NBER Working Paper*, No. 24245, http://dx.doi.org/10.3386/w24245. [105]

UK (2017), *UK Digital Strategy*, Government of the United Kingdom, https://www.gov.uk/government/publications/uk-digital-strategy/uk-digital-strategy. [70]

UK (2017), *UK Industrial Strategy: A Leading Destination to Invest and Grow*, Great Britain and Northern Ireland, http://htps://assets.publishing.service.gov.uk/government/uploads/system/uploads/attachment_data/file/668161/uk-industrial-strategy-international-brochure.pdf. [99]

UNI (2018), *10 Principles for Workers' Data Rights and Privacy*, UNI Global Union, http://www.thefutureworldofwork.org/docs/10-principles-for-workers-data-rights-and-privacy/. [95]

Varian, H. (2018), "Artificial intelligence, economics and industrial organization", *NBER Working Paper*, No. 24839, http://dx.doi.org/10.3386/w24839. [65]

Wachter, S., B. Mittelstadt and L. Floridi (2017), "Transparent, explainable and accountable AI for robotics", *Science Robotics*, 31 May, http://robotics.sciencemag.org/content/2/6/eaan6080. [47]

Wachter, S., B. Mittelstadt and C. Russell (2017), "Counterfactual explanations without opening the black box: Automated decisions and the GDPR", *arXiv* 00399, https://arxiv.org/pdf/1711.00399.pdf. [27]

Weinberger, D. (2018), "Optimization over explanation - Maximizing the benefits of machine learning without sacrificing its intelligence", *Medium*, 28 January, https://medium.com/@dweinberger/optimization-over-explanation-maximizing-the-benefits-we-want-from-machine-learning-without-347ccd9f3a66. [1]

Weinberger, D. (2018), *Playing with AI Fairness*, Google PAIR, 17 September, https://pair-code.github.io/what-if-tool/ai-fairness.html. [22]

Winick, E. (2018), "Every study we could find on what automation will do to jobs, in one chart", *MIT Technology Review*, 25 January, https://www.technologyreview.com/s/610005/every-study-we-could-find-on-what-automation-will-do-to-jobs-in-one-chart/. [86]

Wong, Q. (2017), "At LinkedIn, artificial intelligence is like 'oxygen'", *Mercury News*, 1 June, http://www.mercurynews.com/2017/01/06/at-linkedin-artificial-intelligence-is-like-oxygen. [94]

Yona, G. (2017), "A gentle introduction to the discussion on algorithmic fairness", *Toward Data Science,* 5 October, https://towardsdatascience.com/a-gentle-introduction-to-the-discussion-on-algorithmic-fairness-740bbb469b6.    [18]

Zeng, M. (2018), "Alibaba and the future of business", *Harvard Business Review,* September-October, https://hbr.org/2018/09/alibaba-and-the-future-of-business.    [73]

Notes

[1] More information at https://www.microsoft.com/en-us/ai/ai-for-good.

[2] See https://deepmind.com/applied/deepmind-ethics-society/.

[3] See https://www.blog.google/technology/ai/ai-principles/.

[4] See https://ai.google/static/documents/perspectives-on-issues-in-ai-governance.pdf.

[5] The International Labour Organization, the OECD Guidelines for Multinational Enterprises, or the United Nations Guiding Principles on Business and Human Rights are examples.

[6] More discussions on the topic can be found at https://www.dudumimran.com/2018/05/speaking-about-ai-and-cyber-security-at-the-oecd-forum-2018.html and https://maliciousaireport.com/.

[7] Pierre Chalançon, Chair of the OECD Business and Industry Advisory Committee Consumer Task Force and Vice President Regulatory Affairs, Vorwerk & Co KG, Representation to the European Union – *Science-Fiction is not a Sound Basis for Legislation.*

[8] Among others, this technique has been used to train autonomous vehicles in doing complex manoeuvres, to train the AlphaGo programme and to treat cancer patients; determining the smallest doses and administration frequency that still shrink brain tumours (Matheson, 2018[107]).

[9] Results from a recent study suggest that in many cases synthetic data can successfully replace real data and help scientists overcome privacy constraints (Patki, Wedge and Veeramachaneni, 2016[106]). The authors show that in 70% of the cases the results generated using synthetic data were not significantly different than those obtained using real data.

[10] Schemes such as fully homomorphic encryption, in combination with neural networks, have been successfully tested and employed in this regard (Dowlin, 2016[60]).

[11] See https://www.wipo.int/about-ip/en/artificial_intelligence/ and https://www.uspto.gov/about-us/events/artificial-intelligence-intellectual-property-policy-considerations.

[12] See https://www.ibm.com/watson/stories/creditmutuel/.

[13] As one example, Alibaba no longer employs temporary workers to handle consumer inquiries on days of high volume or special promotions. During Alibaba's biggest sales day in 2017, the chatbot handled more than 95% of customer questions, responding to some 3.5 million consumers (Zeng, 2018[73]). As chatbots take over customer service functions, the role of human customer representatives is evolving to deal with more complicated or personal issues.

# 5. AI policies and initiatives

*Artificial intelligence (AI) policies and initiatives are gaining momentum in governments, companies, technical organisations, civil society and trade unions. Intergovernmental initiatives on AI are also emerging. This chapter collects AI policies, initiatives and strategies from different stakeholders at both national and international levels around the world. It finds that, in general, national government initiatives focus on using AI to improve productivity and competitiveness with actions plans to strengthen: i) factor conditions such as AI research capability; ii) demand conditions; iii) related and supporting industries; iv) firm strategy, structure and competition; as well as v) domestic governance and co-ordination. International initiatives include the OECD* Recommendation of the Council on Artificial Intelligence, *which represents the first intergovernmental policy guidelines for AI and identifies principles and policy priorities for responsible stewardship of trustworthy AI.*

Artificial intelligence (AI) is an increasing priority on the policy agendas for governmental institutions, at both national and international levels. Many national government initiatives to date focus on using AI for productivity and competitiveness. The priorities underlined in national AI action plans can be categorised into five main themes, some of which match Porter's economic competitiveness framework. These priorities are : i) factor conditions such as AI research capability, including skills; ii) demand conditions; iii) related and supporting industries; iv) firm strategy, structure and competition; and v) attention to domestic governance and co-ordination (Box 5.1). In addition, policy consideration for AI issues such as transparency, human rights and ethics is growing.

Among OECD countries and partner economies, Canada, the People's Republic of China (hereafter "China"), France, Germany, India, Sweden, the United Kingdom and the United States have targeted AI strategies. Some countries like Denmark, Japan and Korea include AI-related actions within broader plans. Many other countries – including Australia, Estonia, Finland, Israel, Italy and Spain – are developing a strategy. All strategies aim to increase AI researchers and skilled graduates; to strengthen national AI research capacity; and to translate AI research into public- and private-sector applications. In considering the economic, social, ethical, policy and legal implications of AI advances, the national initiatives reflect differences in national cultures, legal systems, country size and level of AI adoption, although policy implementation is at an early stage. This chapter also examines recent developments in regulations and policies related to AI; however, it does not analyse or assess the realisation of the aims and goals of national initiatives, or the success of different approaches.

AI is also being discussed at international venues such as the Group of Seven (G7), Group of Twenty (G20), OECD, European Union and the United Nations. The European Commission emphasises AI-driven efficiency and flexibility, interaction and co-operation, productivity, competitiveness and growth and quality of citizens' lives. Following the G7 ICT Ministers' Meeting in Japan in April 2016, the G7 ICT and Industry Ministers' Meeting in Turin, Italy in September 2017 shared a vision of "human-centred" AI. It decided to encourage international co-operation and multi-stakeholder dialogue on AI and advance understanding of AI co-operation, supported by the OECD. Ongoing G20 attention to AI can also be noted, particularly with a focus on AI proposed by Japan under its upcoming G20 presidency in 2019 (G20, 2018[1]).

## Principles for AI in society

Several stakeholder groups are actively engaged in discussions on how to steer AI development and deployment to serve all of society. For example, the Institute for Electrical and Electronics Engineers (IEEE) launched its Global Initiative on Ethics of Autonomous and Intelligent Systems in April 2016. It published version 2 of its Ethically Aligned Design principles in December 2017. The final version was planned for early 2019. The Partnership on Artificial Intelligence to Benefit People and Society, launched in September 2016 with a set of tenets, has begun work to develop principles for specific issues such as safety. The Asilomar AI Principles are a set of research, ethics and values for the safe and socially beneficial development of AI in the near and longer term. The AI Initiative brings together experts, practitioners and citizens globally to build common understanding of concepts such as AI explainability.

**Table 5.1. Selection of sets of guidelines for AI developed by stakeholders (non-exhaustive)**

| Reference | Guidelines for AI developed by stakeholders |
|---|---|
| ACM | ACM (2017), "2018 ACM Code of Ethics and Professional Conduct: Draft 3", Association for Computing Machinery Committee on Professional Ethics, https://ethics.acm.org/2018-code-draft-3/<br>USACM (2017), "Statement on Algorithmic Transparency and Accountability", Association for Computing Machinery US Public Policy Council, www.acm.org/binaries/content/assets/public-policy/2017_usacm_statement_algorithms.pdf |
| AI Safety | Amodei, D. et al. (2016), "Concrete Problems in AI Safety", 25 July, https://arxiv.org/pdf/1606.06565.pdf |
| Asilomar | FLI (2017), "Asilomar AI Principles", Future of Life Institute, https://futureoflife.org/ai-principles/ |
| COMEST | COMEST (2017), "Report of COMEST on Robotics Ethics", World Commission on the Ethics of Scientific Knowledge and Technology, http://unesdoc.unesco.org/images/0025/002539/253952E.pdf |
| Economou | Economou, N. (2017) "A 'principled' artificial intelligence could improve justice", 3 October, *Aba Journal* www.abajournal.com/legalrebels/article/a_principled_artificial_intelligence_could_improve_justice |
| EGE | EGE (2018), "Statement on Artificial Intelligence, Robotics and Autonomous Systems", European Group on Ethics in Science and New Technologies, http://ec.europa.eu/research/ege/pdf/ege_ai_statement_2018.pdf |
| EPSRC | EPSRC (2010), "Principles of Robotics" , Engineering and Physical Sciences Research Council, https://epsrc.ukri.org/research/ourportfolio/themes/engineering/activities/principlesofrobotics/ |
| FATML | FATML (2016), "Principles for Accountable Algorithms and a Social Impact Statement for Algorithms", Fairness, Accountability, and Transparency in Machine Learning, www.fatml.org/resources/principles-for-accountable-algorithms |
| FPF | FPF (2018), "Beyond Explainability: A Practical Guide to Managing Risk in Machine Learning Models", The Future of Privacy Forum, https://fpf.org/wp-content/uploads/2018/06/Beyond-Explainability.pdf |
| Google | Google (2018), "AI at Google: Our Principles", https://www.blog.google/technology/ai/ai-principles/ |
| IEEE | IEEE (2017), Global Initiative on Ethics of Autonomous and Intelligent Systems, "Ethically Aligned Design Version 2", Institute of Electrical and Electronics Engineers, http://standards.ieee.org/develop/indconn/ec/ead_v2.pdf |
| Intel | Intel (2017), "AI - The Public Policy Opportunity", https://blogs.intel.com/policy/files/2017/10/Intel-Artificial-Intelligence-Public-Policy-White-Paper-2017.pdf |
| ITI | ITI (2017), "AI Policy Principles", Information Technology Industry Council, www.itic.org/resources/AI-Policy-Principles-FullReport2.pdf |
| JSAI | JSAI (2017), "The Japanese Society for Artificial Intelligence Ethical Guidelines", The Japanese Society for Artificial Intelligence, http://ai-elsi.org/wp-content/uploads/2017/05/JSAI-Ethical-Guidelines-1.pdf |
| MIC | MIC (2017), "Draft AI R&D Guidelines for International Discussions", Japanese Ministry of Internal Affairs and Communication, www.soumu.go.jp/main_content/000507517.pdf |
| MIC | MIC (2018), "Draft AI Utilization Principles", Japanese Ministry of Internal Affairs and Communication, www.soumu.go.jp/main_content/000581310.pdf |
| Montreal | UoM (2017), "The Montreal Declaration for a Responsible Development of Artificial Intelligence", University of Montreal, www.montrealdeclaration-responsibleai.com/ |
| Nadella | Nadella, S. (2016) "The Partnership of the Future", 28 June, Slate, www.slate.com/articles/technology/future_tense/2016/06/microsoft_ceo_satya_nadella_humans_and_a_i_can_work_together_to_solve_society.html |
| PAI | PAI (2016), "TENETS", Partnership on AI, www.partnershiponai.org/tenets/ |
| Polonski | Polonski, V. (2018) "The Hard Problem of AI Ethics - Three Guidelines for Building Morality Into Machines" , 28 February, Forum Network on Digitalisation and Trust, www.oecd-forum.org/users/80891-dr-vyacheslav-polonski/posts/30743-the-hard-problem-of-ai-ethics-three-guidelines-for-building-morality-into-machines |
| Taddeo and Floridi | Taddeo, M. and L. Floridi (2018), "How AI can be a force for good", *Science*, 24 August, Vol. 61/6404, pp. 751-752, http://science.sciencemag.org/content/361/6404/751 |
| The Public Voice Coalition | UGAI (2018), "Universal Guidelines on Artificial Intelligence", The Public Voice Coalition, https://thepublicvoice.org/ai-universal-guidelines/ |
| Tokyo Statement | Next Generation Artificial Intelligence Research Center (2017), "The Tokyo Statement – Co-operation for Beneficial AI", www.ai.u-tokyo.ac.jp/tokyo-statement.html |
| Twomey | Twomey, P. (2018), "Toward a G20 Framework for Artificial Intelligence in the Workplace" , *CIGI Papers*, No 178, Centre for International Governance Innovation, www.cigionline.org/sites/default/files/documents/Paper%20No.178.pdf |
| UNI | UNI Global Union (2017), "Top 10 Principles for Ethical Artificial Intelligence", www.thefutureworldofwork.org/media/35420/uni_ethical_ai.pdf |

Several initiatives have developed valuable sets of principles to guide AI development (Table 5.1), many of which focus on the technical communities that conduct research and development (R&D) of AI systems. Many of these principles were developed in multi-stakeholder processes. However, they can be categorised broadly into five communities:

technical, private sector, government, academic and labour. The technical community includes the Future of Life Institute; the IEEE; the Japanese Society for Artificial Intelligence; Fairness, Accountability and Transparency in Machine Learning; and the Association for Computing Machinery. Examples of a private-sector focus include the Partnership on AI; the Information Industry Technology Council; and Satya Nadella, chief executive officer of Microsoft. The government focus includes the Japanese Ministry of Internal Affairs and Communications; the World Commission on the Ethics of Science and Technology; and the Engineering and Physical Sciences Research Council. Examples of the academic focus include the Université de Montréal and Nicolas Economou, the chief executive officer of H5 and special advisor on the AI Initiative of the Future Society at Harvard Kennedy School. UNI Global Union is an example of the labour community.

Common themes emerge from initiatives. Different stakeholders have developed guidelines, including human values and rights; non-discrimination; awareness and control; access to data; privacy and control; safety and security; skills; transparency and explainability; accountability and responsibility; whole of society dialogue; and measurement.

In May 2018, the OECD's Committee on Digital Economy Policy established the AI Group of Experts at the OECD (AIGO), with the aim of scoping principles for public policy and international co-operation that would foster trust in and adoption of AI (OECD, 2019[2]). This work informed the development of the OECD *Recommendation of the Council on Artificial Intelligence* (OECD, 2019[3]), to which 42 national governments adhered on 22 May 2019.

## National initiatives

### *Overview of AI national policy responses*

Many countries have announced national AI strategies and policy initiatives, which commonly aim to ensure a leadership position in AI. The strategies and initiatives set objectives and targets that require concerted action by all stakeholders. Governments' role is often as a convener and facilitator. Box 5.1 provides elements often seen in policies and measures for fostering national competitiveness in AI. In addition, some countries have created, or assigned responsibility to, a specific public entity for AI and data ethics issues.

---

**Box 5.1. How do countries seek to develop competitive advantage in AI?**

Porter identified four determinants to gain national competitive advantage in a specific industry: i) factor conditions; ii) demand conditions; iii) related and supporting industries; and iv) firm strategy, structure and competition. Porter acknowledged that companies are the actors that create competitive advantage in industries. However, he emphasised the key role of governments in supporting and enabling the four determinants of the national industrial development processes.

- **Factor conditions**: This determinant depends on geography, availability of skilled labour, level of education and research capabilities. Countries are strengthening AI research capability through different measures that include: i) creating AI research institutes; ii) creating new AI-related graduate and doctoral degrees at universities, and adjusting existing degrees to include AI courses, e.g. in scientific disciplines; and iii) attracting domestic and foreign talent e.g. by increasing visas for AI experts.

---

---

**Box 5.1 How do countries seek to develop competitive advantage in AI? (*cont.*)**

- **Demand conditions**: Several countries identify strategic sectors for AI development, notably transportation, healthcare and public services. They are putting in place measures to encourage domestic consumer demand for AI services in these specific industries. In public services, some governments are ensuring that AI systems meet certain standards, e.g. of accuracy or robustness, through public procurement policies.

- **Related and supporting industries**: AI competitiveness requires access to digital infrastructures and services, data, computing power and broadband connectivity. A number of countries are planning AI-focused technology clusters and support structures for small and medium-sized enterprises (SMEs).

- **Firm strategy, structure and competition**: Some of the approaches that countries are taking to foster private investment and competition in AI include: i) preparing AI development roadmaps for fostering private investment; ii) encouraging international AI companies to invest domestically, e.g. by opening AI laboratories; and iii) experimenting with policy approaches such as regulatory sandboxes for AI applications to encourage firms to innovate.

In addition, for the effective implementation of national AI initiatives, many countries are considering appropriate governance mechanisms to ensure a co-ordinated, whole-of-government approach. For example, France has established an AI co-ordination function within the Prime Minister's Office to implement the French AI strategy.

*Source*: Porter (1990[4]), "The competitive advantage of nations", https://hbr.org/1990/03/the-competitive-advantage-of-nations.

---

### *Argentina*

In July 2019, the Argentine government planned to release a ten-year National AI Strategy. This followed an assessment phase in 2018 by the Ministry of Science, Technology and Productive Innovation as part of the Argentina Digital Agenda 2030 and Argentina's Innovative Plan 2030. Thematic priorities for the National AI Strategy include: talent and education, data, R&D and innovation, supercomputing infrastructure, actions to facilitate job transitions, and facilitating public-private co-operation on data use. Thematic priorities also include public services and manufacturing (as target sectors for AI development). The strategy's cross-cutting themes are: i) investment, ethics and regulation; ii) communication and awareness building; and iii) international co-operation.

The strategy involves seven ministries and envisions the development of a national AI Innovation Hub to implement projects in each thematic group. Each thematic priority will have an expert steering group charged with defining goals and metrics to measure progress.

### *Australia*

The Australian government planned over AUD 28 million (USD 21 million) in its 2018/19 budget to build capability in, and support the responsible development of, AI in Australia. This budget will finance the following projects: i) Co-operative Research Centre projects with a specific focus on AI (USD 18 million); ii) AI-focused PhD scholarships (USD 1 million); iii) development of online resources to teach AI in schools (USD 1.1 million); iv) development of an AI technology roadmap exploring AI's impacts on industries, workforce opportunities and challenges and implications for education and training (USD 250 000); v) development of an AI ethics framework using a case study approach (USD 367 000); and vi) development of an AI standards roadmap (USD 72 000 with matching funds from industry).

The Department of Industry, Innovation and Science is also undertaking AI-related projects. They include tasking the Australian Council of Learned Academies with exploring the opportunities, risks and consequences for Australia of broad uptake of AI over the next decade. The Australian Human Rights Commission also launched a major project in July 2018 on the relationship between human rights and technology. It includes an issues paper and international conference; a final report was planned for 2019/20.

## Brazil

Brazil's digital transformation strategy E-Digital of March 2018 harmonises and co-ordinates different governmental initiatives on digital issues to advance the Sustainable Development Goals in Brazil. On AI specifically, E-Digital includes action "to evaluate potential economic and social impact of (…) artificial intelligence and big data, and to propose policies that mitigate their negative effects and maximise positive results" (Brazil, 2018[5]). E-Digital also prioritises the allocation of resources towards AI research, development and innovation (RD&I), and capacity building. Brazil plans to launch a specific AI strategy in 2019. It is actively involved in international discussions on AI technical standardisation and policy.

From 2014 to early 2019, the Ministry of Science, Technology, Innovation and Communication has provided incentives and financial support for 16 different projects on AI, and to 59 AI start-ups. In addition, 39 initiatives use AI in e-government at the federal level. These initiatives aim, notably, to improve administrative and assessment procedures, e.g. in social services, citizen services, jobs advertising, etc. A new institute for AI research – the Artificial Intelligence Advanced Institute – was set up in 2019. It promotes partnerships between universities and companies on joint AI RD&I projects. Specifically, it targets fields such as agriculture, smart cities, digital governance, infrastructure, environment, natural resources, and security and defence.

## Canada

Canada is seeking to position itself as an AI leader notably with the Pan-Canadian AI Strategy launched in March 2017 (CIFAR, 2017[6]). The strategy is led by the non-profit Canadian Institute for Advanced Research and backed with government funding of CAD 125 million (USD 100 million). Over five years, the funds will support programmes to expand Canada's human capital, support AI research in Canada and translate AI research into public- and private-sector applications. The goals of the Pan-Canadian AI Strategy are:

1. Increase AI researchers and skilled graduates in Canada.

2. Establish interconnected nodes of scientific excellence in Canada's three major AI institutes: in Edmonton (Alberta Machine Intelligence Institute), Montreal (Montreal Institute for Learning Algorithms) and Toronto (Vector Institute for Artificial Intelligence).

3. Develop a global programme on AI in Society and global thought leadership on the economic, social, ethical, policy and legal implications of advances in AI.

4. Support a national research community on AI.

The federal government through the National Research Council Canada (NRC) plans research investments totalling CAD 50 million (USD 40 million) over a seven-year period to apply

AI to key programme themes including: data analytics, AI for Design, cyber security, Canadian indigenous languages, support for federal superclusters and collaboration centres with Canadian universities, as well as strategic partnerships with international partners.

In addition to this federal funding, the Quebec government is allocating CAD 100 million (USD 80 million) to the AI community in Montreal; Ontario is providing CAD 50 million (USD 40 million) to the Vector Institute for Artificial Intelligence. In 2016, the Canada First Research Excellence Fund allocated CAD 93.6 million (USD 75 million) to three universities for cutting-edge research in deep learning: the Université de Montréal, Polytechnique Montréal and HEC Montréal. Facebook and other dynamic private companies like ElementAI are active in Canada.

The Quebec government plans to create a world observatory on the social impacts of AI and digital technologies (Fonds de recherche du Québec, 2018[7]). A workshop in March 2018 began to consider the observatory's mandate and potential model, governance mode, funding and international co-operation, as well as sectors and issues of focus. The Quebec government has allocated CAD 5 million (USD 3.7 million) to help implement the observatory.

Canada is also working with partners internationally to advance AI initiatives. For example, the governments of Canada and France announced in July 2018 that they would work together to establish a new International Panel on AI. This Panel's mission will be to support and guide the responsible adoption of AI that is human-centred and grounded in human rights, inclusion, diversity, innovation and economic growth.

## *China*

In May 2016, the Chinese government published a three-year national AI plan formulated jointly by the National Development and Reform Commission, the Ministry of Science and Technology, the Ministry of Industry and Information Technology and the Cyberspace Administration of China. AI was included in the Internet Plus initiative, which was established in 2015 as a national strategy to spur economic growth driven by innovative, Internet-related technologies in the period 2016-18 (Jing and Dai, 2018[8]). It focuses on: i) enhancing AI hardware capacity; ii) strong platform ecosystems; iii) AI applications in important socio-economic areas; and iv) AI's impact on society. In it, the Chinese government envisioned creating a USD 15 billion market by 2018 through R&D for the Chinese AI industry (China, 2016[9]).

Mid-2017, China's State Council released the *Guideline on Next Generation AI Development Plan*, which provides China's long-term perspective on AI with industrial goals for each period. These comprise: i) AI-driven economic growth in China by 2020; ii) major breakthroughs in basic theories by 2025 and in building an intelligent society; and iii) for China to be a global AI innovation centre by 2030 and to build up an AI industry of RMB 1 trillion (USD 150 billion) (China, 2017[10]). The plan's implementation seems to be advancing throughout government and China has been developing leadership in AI with state support and private company dynamism. China's State Council set objectives for "new-generation information technology" as a strategic industry targeted to account for 15% of gross domestic product by 2020.

In its 13th Five-Year Plan timeframe (2016-20), China ambitions to transform itself into a science and technology leader, with 16 "Science and Technology Innovation 2030 Megaprojects", including "AI 2.0". The plan has provided impetus for action in the public sector (Kania, 2018[11]). The plan asks companies to accelerate AI hardware and software R&D, including in AI-based vision, voice and biometric recognition, human-machine interfaces and smart controls.

On 18 January 2018, China established a national AI standardisation group and a national AI expert advisory group. At the same time, the National Standardisation Management Committee Second Ministry of Industry released a white paper on AI standardisation. The paper was supported by the China Electronic Standardisation Institute (a division in the Ministry of Industry and Information Technology) (China, 2018[12]).

Private Chinese companies' attention to AI predate the more recent government support. Chinese companies such as Baidu, Alibaba and Tencent have made significant efforts and investments in AI. Chinese industry has focused on applications and data integration, while the central government focuses on basic algorithms, open data and conceptual work. City governments focus on the use of applications and open data at a municipal level.

### Czech Republic

The Czech government called for a study on AI implementation in 2018 to develop strategic goals and support negotiations at European and international levels. A team of academic experts from the Technology Centre of the Academy of Science of the Czech Republic, the Czech technological university in Prague and the Czech Academy of Sciences' Institute of State and Law submitted a report called *Analysis of the Development Potential of Artificial Intelligence in the Czech Republic* (OGCR, 2018[13]). This reports maps: i) the current state of AI implementation in the Czech Republic; ii) the potential impact of AI on the Czech labour market; and iii) ethical, legal and regulatory aspects of AI development in the country.

### Denmark

Denmark published the *Strategy for Denmark's Digital Growth* in January 2018. It aims for Denmark to become a digital frontrunner, with all Danes benefiting from the transformation. The strategy introduces initiatives to seize growth opportunities from AI, big data and Internet of Things (IoT) technologies. Its strategic focus includes: i) creating a digital hub for public-private partnerships; ii) assisting SMEs with data-driven business development and digitalisation; iii) establishing educational institutions in a Technology Pact to foster technical and digital skills; iv) strengthening cyber security in companies; and v) developing agile regulation to facilitate new business models and experimentation. The Danish government has committed DKK 1 billion (USD 160 million) until 2025 to implement the strategy. It divides this commitment into DKK 75 million (USD 12 million) for 2018, and DKK 125 million (USD 20 million) in 2019-25. The largest part of this budget will be allocated to skills development initiatives, followed by the digital hub creation and support for SMEs (Denmark, 2018[14]).

### Estonia

Estonia is planning the next step of its e-governance system powered by AI to save costs and improve efficiency. It is also experimenting with e-healthcare and situational awareness. Estonia aims to improve lives and cities and support human values. On the enforcement side, Estonia targets the core values of ethics, liability, integrity and accountability. This is in place of a focus on rapidly evolving technology and building an enforcement system based on blockchain that mitigates integrity and accountability risks. A pilot project is planned in 2018.

With StreetLEGAL, self-driving cars have been tested on Estonian public roads since March 2017. Estonia is also the first government discussing the legalisation of AI. This would entail giving representative rights, and responsibilities, to algorithms to buy and sell services on their owners' behalf. In 2016, the Estonian government created a task force to look into the problem of accountability in ML algorithms and the need for legislation of AI, together with the Ministry of Economic Affairs and Communications and the Government Office (Kaevats, 2017[15]; Kaevats, 25 September 2017[16]).

## *Finland*

Finland aims to develop a safe and democratic society with AI; to provide the best public services in the world; and for AI to bring new prosperity, growth and productivity to citizens. The AI strategy *Finland's Age of Artificial Intelligence*, published in October 2017, is a roadmap for the country to leverage its educated population, advanced digitalisation and public sector data resources. At the same time, the strategy foresees building international links in research and investment, and encouraging private investments. Finland hopes to double its national economic growth by 2035 thanks to AI. Eight key actions for AI-enabled growth, productivity and well-being are: i) enhancing companies' competitiveness; ii) using data in all sectors; iii) speeding up and simplifying AI adoption; iv) ensuring top-level expertise; v) making bold decisions and investments; vi) building the world's best public services; vii) establishing new co-operation models; and viii) making Finland a trendsetter in the age of AI. The report highlights using AI to improve public services. For example, the Finnish Immigration Service uses the national customer service robot network called Aurora to provide multilingual communication (Finland, 2017[17]).

In February 2018, the government also created a funding entity for AI research and commercial projects. The entity will allocate EUR 200 million (USD 235 million) in grants and incentives to the private sector, including SMEs. Finland reports some 250 companies working on AI development. This includes professionals and patients in the Finnish healthcare industry organisations and healthcare system, and the associated in-depth reforms (Sivonen, 2017[18]). The role of the Finnish state-funded Technical Research Centre of Finland and the Finnish Funding Agency for Technology and Innovation will also be expanded.

## *France*

French President Emmanuel Macron announced France's AI strategy on 29 March 2018. It allocates EUR 1.5 billion of public funding into AI by 2022 to help France become an AI research and innovation leader. The measures are largely based on recommendations in the report developed by a member of parliament, Cédric Villani (Villani, 2018[19]). The strategy calls for investing in public research and education, building world-class research hubs linked to industry through public-private partnerships, and attracting foreign and French elite AI researchers working abroad. To develop the AI ecosystem in France, the strategy's approach is to "upgrade" existing industries. Starting from applications in health, environment, transport and defence, it aims to help use AI to renew existing industries. It proposes to prioritise access to data by creating "data commons" between private and public actors; adapting copyright law to facilitate data mining; and opening public sector data such as health to industry partners.

The strategy also outlines initial plans for AI-induced disruptions, taking a firm stance on data transfers out of Europe (Thompson, 2018[20]). It would create a central data agency with a team of about 30 advisory experts on AI applications across government. The ethical and philosophical boundaries articulated in the strategy include algorithm transparency as a core principle. Algorithms developed by the French government or with public funding, for example, will reportedly be open. Respect for privacy and other human rights will be "by design". The strategy also develops vocational training in professions threatened by AI. It calls for policy experimentation in the labour market and for dialogue on how to share AI-generated value added across the value chain. A French report on AI and work was also released in late March (Benhamou and Janin, 2018[21]).

## Germany

The German federal government launched its AI strategy in December 2018 (Germany, 2018[22]). Germany aims to become a leading centre for AI by pursuing speedy and comprehensive transfer of research findings into applications, with "AI made in Germany" becoming a strong export and a globally recognised quality mark. Measures to achieve this goal include new research centres, enhanced Franco-Germany research co-operation, funding for cluster development and support for SMEs. It also addresses infrastructure requirements, enhanced access to data, skills development, security to prevent misuse and ethical dimensions.

In June 2017, the federal Ministry of Transport and Digital Infrastructure developed ethical guidelines for self-driving cars. The guidelines, developed by the Ethics Commission of the ministry, stipulate 15 rules for programmed decisions embedded in self-driving cars. The commission considered ethical questions in depth, including whose life to prioritise (known as the "Trolley problem"). The guidelines provide that self-driving cars should be programmed to consider all human lives as equal. If a choice is needed between people, self-driving cars should choose to hit whichever person would be hurt less, regardless of age, race or gender. The commission also makes clear that no obligation should be imposed on individuals to sacrifice themselves for others (Germany, 2017[23]).

## Hungary

Hungary created an AI Coalition in October 2018 as a partnership between state agencies, leading IT businesses and universities. The coalition is formulating an AI strategy to establish Hungary as an AI innovator and researching the social and economic impacts of AI on society. It includes some 70 academic research centres, businesses and state agencies and is viewed as a forum for cross-sector co-operation in AI R&D. The Budapest Technology and Economics and Eötvös Loránd Universities are part of the consortium investing EUR 20 million (USD 23.5 million) in AI4EU, a project to develop an AI-on-demand-platform in Europe.

## India

India published its AI strategy in June 2018. The strategy provides recommendations for India to become a leading nation in AI by empowering human capability and ensuring social and inclusive growth. The report named its inclusive approach #AIFORALL. The strategy identifies the following strategic focus areas for AI applications: healthcare, agriculture, education, smart cities and transportation. Low research capability and lack of data ecosystems in India are identified as challenges to realise the full potential of AI. The strategy makes several recommendations. India should create two-tiered research institutes (for both basic and applied research). It needs to set up learning platforms for the current workforce. The country should also create targeted data sets and incubation hubs for start-ups. Finally, it should establish a regulatory framework for data protection and cyber security (India, 2018[24]).

## Italy

Italy published a white paper "Artificial Intelligence at the Service of Citizens" in March 2018 that was developed by a task force of the Agency for Digital Italy. It focused on how the public administration can leverage AI technologies to serve people and business, and increase public-service efficiency and user satisfaction. The report identified challenges to implement AI in public services related to ethics, technology, data availability and impact measurement. The report includes recommendations on issues such as promoting a national platform for labelled data, algorithms and learning models; developing skills; and creating a National Competence Centre and a Trans-disciplinary Centre on AI. The report also called for guidelines and processes to increase levels of control and facilitate data sharing among all European countries on cyberattacks to and from AI (Italy, 2018[25]).

## *Japan*

The Japanese Cabinet Office established a Strategic Council for AI Technology in April 2016 to promote AI technology R&D and business applications. The council published an *Artificial Intelligence Technology Strategy* in March 2017 that identified critical issues. These included the need to increase investment, facilitate use and access to data, and increase the numbers of AI researchers and engineers. The strategy also identified strategic areas in which AI could bring significant benefits: productivity; health, medical care and welfare; mobility; and information security (Japan, 2017[26]).

Japan's *Integrated Innovation Strategy*, published by the Cabinet Office in June 2018, has a set of AI policy actions (Japan, 2018[27]). The strategy included convening multi-stakeholder discussions on ethical, legal and societal issues of AI. This resulted in the Cabinet Office publishing *Social Principles for Human-centric AI* in April 2019 (Japan, 2019[28]).

At the G7 ICT Ministerial meeting in Takamatsu in April 2016, Japan proposed the formulation of shared principles for AI R&D. A group of experts called Conference toward AI Network Society developed *Draft AI R&D Guidelines for International Discussions*, which were published by the Japanese Ministry of Internal Affairs and Communications in July 2017. These guidelines seek primarily to balance benefits and risks of AI networks, while ensuring technological neutrality and avoiding excessive burden on developers. The guidelines consist of nine principles that researchers and developers of AI systems should consider (Japan, 2017[29]). Table 5.2 provides the abstract of the guidelines. The Conference published the *Draft AI Utilization Principles* in July 2018 as an outcome of the discussion (Japan, 2018[30]).

**Table 5.2. R&D Principles provided in the AI R&D Guidelines**

| Principle of: | Developers should: |
| --- | --- |
| I. Collaboration | Pay attention to the interconnectivity and interoperability of AI systems. |
| II. Transparency | Pay attention to the verifiability of inputs/outputs of AI systems and explainability of their decisions. |
| III. Controllability | Pay attention to the controllability of AI systems. |
| IV. Safety | Ensure that AI systems do not harm the life, body or property of users or third parties through actuators or other devices. |
| V. Security | Pay attention to the security of AI systems. |
| VI. Privacy | Take into consideration that AI systems will not infringe the privacy of users or third parties. |
| VII. Ethics | Respect human dignity and individual autonomy in R&D of AI systems. |
| VIII. User assistance | Take into consideration that AI systems will support users and make it possible to give them opportunities for choice in appropriate manners. |
| IX. Accountability | Make efforts to fulfil their accountability to stakeholders, including users of AI systems. |

*Source*: Japan (2017[29]), *Draft AI R&D Guidelines for International Discussions*, www.soumu.go.jp/main_content/000507517.pdf.

## *Korea*

The Korean government published the *Intelligent Information Industry Development Strategy* in March 2016. It announced public investment of KRW 1 trillion (USD 940 million) by 2020 in the field of AI and related information technologies such as IoT and cloud computing. This strategy aimed to create a new intelligent information industry ecosystem and to encourage KRW 2.5 trillion (USD 2.3 billion) of private investment by 2020. Under the strategy, the government has three goals. First, it plans to launch AI development flagship projects, for example in the areas of language-visual-space-emotional intelligence technology. Second, it seeks to strengthen AI-related workforce skills. Third, it will promote access and use of data by government, companies and research institutes (Korea, 2016[31]).

In December 2016, the Korean government published the *Mid- to Long-Term Master Plan in Preparation for the Intelligence Information Society*. The plan contains national policies to respond to the changes and challenges of the 4th Industrial Revolution. To achieve its vision of a "human-centric intelligent society", it aims to establish the foundations for world-class intelligent IT. Such IT could be applied across industries and be used to upgrade social policies. To implement the plan, the government is creating large-scale test beds to help develop new services and products, including better public services (Korea, 2016[32]).

In May 2018, the Korean government released a national plan to invest KRW 2.2 trillion (USD 2 billion) by 2022 to strengthen its AI R&D capability. The plan would create six AI research institutes; develop AI talent through 4 500 AI scholarships and short-term intensive training courses; and accelerate development of AI chips (Peng, 2018[33]).

## *Mexico*

In Mexico, the National Council of Science and Technology created an Artificial Intelligence Research Centre in 2004, which leads the development of intelligent systems.

In June 2018, a white paper entitled "Towards an AI Strategy in Mexico: Harnessing the AI Revolution" was published.[1] The report finds that Mexico ranks 22 out of 35 OECD countries in Oxford Insight's "AI Readiness Index". Its composite score was derived from averaging nine metrics ranging from digital skills to government innovation. Mexico scores well for its open data policies and digital infrastructure, but poorly in areas such as technical skills, digitalisation and public sector innovation. The report recommends policy actions to further develop and deploy AI in Mexico. These recommendations span five areas of government: government and public services; R&D; capacity, skills and education; data and digital infrastructure; and ethics and regulation (Mexico, 2018[34]).

## *The Netherlands*

In 2018, through adoption of the National Digitalisation Strategy, the government of the Netherlands made two commitments. First, it will leverage social and economic opportunities. Second, it will strengthen enabling conditions, including skills, data policy, trust and resilience, fundamental rights and ethics (e.g. the influence of algorithms on autonomy and equal treatment) and AI-focused research and innovation. In October 2018, "AI for the Netherlands" (AINED), a Dutch coalition of industry and academia, published goals and actions for a national AI plan. These include focusing on promoting access to AI talent and skills, as well as high-value public data. They also aim to facilitate AI-driven business development and promote large-scale use of AI in government. In addition, they envision creating socio-economic and ethical frameworks for AI, encouraging public-private co-operation in key sectors and value chains, and establishing the Netherlands as a world-class AI research centre. The Dutch government plans to finalise a whole-of-government strategic action plan for AI before mid-2019. It was to consider the AINED report, the EU co-ordinated plan and discussions of the European Commission (EC) High-Level Expert Group on AI (AI HLEG).

## *Norway*

Norway has taken AI initiatives as part of the Digital Agenda for Norway and Long-term Plan for Research and Higher Education that include:

- Creation of several AI labs, such as the Norwegian Open AI-lab at the Norwegian University for Science and Technology. The Open AI-lab is sponsored by several companies. It focuses on energy, maritime, aquaculture, telecom, digital banking, health and biomedicine, where Norway has a strong international position.

- A skills-reform programme called Learning for Life with a proposed 2019 budget of NOK 130 million (USD 16 million) to help the workforce develop or upgrade skills in AI, healthcare and other areas.

- An open data strategy by which government agencies must make their data available in machine-readable formats using application program interfaces and register the available datasets in a common catalogue.

- A platform to develop guidelines and ethical principles governing the use of AI.

- Regulatory reform to allow self-driving vehicles testing on the road – including test-driving without a driver in the vehicle.

## *Russian Federation*

The Russian government developed its digital strategy entitled *Digital Economy of the Russian Federation* in July 2017. The strategy prioritises leveraging AI development, including favourable legal conditions to facilitate R&D activities. It also seeks to provide incentives for state companies to participate in the nationwide research communities (competence centres). In addition, it promotes activities to develop national standards for AI technologies (Russia, 2017[35]). Prior to the digital strategy, the government invested in various AI projects and created instruments to develop public-private partnerships. The Russian AI Association is encouraging co-operation between academia and companies to facilitate technology transfer to companies.

## *Saudi Arabia*

Saudi Arabia announced its Vision 2030 in 2016. It delivers an economic reform plan to stimulate new industries and diversify the economy, facilitate public-private business models and, ultimately, reduce the country's dependence on oil revenues. Vision 2030 views digital transformation as a key means to develop the economy by leveraging data, AI and industrial automation. Priority sectors, including for the launch of innovation centres, include healthcare, government services, sustainable energy and water, manufacturing, and mobility and transportation. The government is drafting its national AI strategy that aims to build an innovative and ethical AI ecosystem in Saudi Arabia by 2030.

Saudi Arabia is developing an enabling ecosystem for AI that includes high speed broadband and 5G deployment, access to data and security. It is also encouraging early adoption of AI concepts and solutions through several smart city projects to catalyse new solutions. These build on the momentum of NEOM, a smart city megaproject launched in 2017 that invests a significant SAR 1.8 trillion (USD 500 billion). Saudi Arabia is also actively involved in global discussions on AI governance frameworks.

## Singapore

Infocomm Media Development Authority published the *Digital Economy Framework for Action* in May 2018. This maps out a framework for action to transform Singapore into a leading digital economy, identifying AI as a frontier technology to drive Singapore's digital transformation (Singapore, 2018[36]).

The Personal Data Protection Commission published a model AI governance framework in January 2019 to promote responsible adoption of AI in Singapore. The model framework provides practical guidance to convert ethical principles into implementable practices. It builds on a discussion paper and national discussions by a Regulators' Roundtable, a community of practice comprising sector regulators and public agencies. Organisations can adopt the model framework voluntarily. It also serves as a basis to develop sector-specific AI governance frameworks.

A multi-stakeholder Advisory Council on the Ethical Use of AI and Data was formed in June 2018. It advises Singapore's government on ethical, legal, regulatory and policy issues arising from the commercial deployment of AI. To that end, it aims to support industry adoption of AI and the accountable and responsible rollout of AI products and services.

To develop Singapore into a leading knowledge centre with international expertise in AI policy and regulations, the country set up a five-year Research Programme on the Governance of AI and Data Use in September 2018. It is headquartered at the Centre for AI & Data Governance of the Singapore Management University School of Law. The Centre focuses on industry-relevant research of AI as it relates to industry, society and business.

## Sweden

In May 2018, the Swedish government published a report called *Artificial Intelligence in Swedish Business and Society*. The report, which aims to strengthen AI research and innovation in Sweden, outlines six strategic priorities: i) industrial development, including manufacturing; ii) travel and transportation; iii) sustainable and smart cities; iv) healthcare; v) financial services; and vi) security, including police and customs. The report highlights the need to reach a critical mass in research, education and innovation. It also calls for co-operation regarding investment for research and infrastructure, education, regulatory development and mobility of labour forces (Vinnova, 2018[37]).

## Turkey

The Scientific and Technological Research Council of Turkey – the leading agency for the management and funding of research in Turkey – has funded numerous AI R&D projects. It plans to open a multilateral call for AI projects in the context of the EUREKA intergovernmental network for innovation. The Turkish Ministry of Science and Technology has developed a national digital roadmap in the context of Turkey's Industrial Digital Transformation Platform. Part of this roadmap focuses on technological advancement in emerging digital technologies such as AI.

## United Kingdom

The *UK Digital Strategy* published in March 2017 recognises AI as key to help grow the United Kingdom's digital economy (UK, 2017[38]). The strategy allocates GBP 17.3 million (USD 22.3 million) in funding for UK universities to develop AI and robotics technologies. The government has increased investment in AI R&D by GBP 4.7 billion (USD 6.6 billion) over the next four years, partly through its Industrial Strategy Challenge Fund.

In October 2017, the government published an industry-led review on the United Kingdom's AI industry. The report identifies the United Kingdom as an international centre of AI expertise, in part as a result of pioneering computer scientists such as Alan Turing. The UK government estimated that AI could add GBP 579.7 billion (USD 814 billion) to the domestic economy. AI tools used in the United Kingdom include a personal health guide (Your.MD), a chatbot developed for bank customers and a platform to help children learn and teachers provide personalised education programmes. The report provided 18 recommendations, such as improving access to data and data sharing by developing Data Trusts. It also recommends improving the AI skills supply through industry-sponsored Masters in AI. Other suggested priorities include maximising AI research by co-ordinating demand for computing capacity for AI research among relevant institutions; supporting uptake of AI through a UK AI Council; and developing a framework to improve transparency and accountability of AI-driven decisions (Hall and Pesenti, 2017[39]).

The UK government published its industrial strategy in November 2017. The strategy identifies AI as one of four "Grand Challenges" to place the United Kingdom at the forefront of industries of the future and ensure it takes advantage of major global changes (UK, 2017[40]). In April 2018, the United Kingdom published the *AI Sector Deal*: a GBP 950 million (USD 1.2 billion) investment package that builds on the United Kingdom's strengths and seeks to maintain a world-class ecosystem. It has three principal areas of focus: skills and talent; the drive for adoption; and data and infrastructure (UK, 2018[41]).

The government established an Office for Artificial Intelligence (OAI) focused on implementing the Sector Deal and driving adoption more widely. It also established a Centre for Data Ethics and Innovation. The Centre focuses on strengthening the governance landscape to enable innovation, while ensuring public confidence. It will supply government with independent, expert advice on measures needed for safe, ethical and ground-breaking innovation in data-driven and AI-based technologies. To that end, it planned to launch a pilot data trust by the end of 2019. An AI Council that draws on industry expertise works closely with the OAI (UK, 2018[42]).

### United States

On 11 February 2019, President Trump signed Executive Order 13859 on Maintaining American Leadership in Artificial Intelligence, launching the American AI Initiative. The Initiative directs actions in five key areas: i) invest in AI R&D; ii) unleash AI resources; iii) set guidance for AI regulation and technical standards; iv) build the AI workforce; and v) engage internationally in support of American AI research and innovation and to open markets for American AI industries.

The Initiative is the culmination of a series of Administration actions to accelerate American leadership in AI. The White House hosted the first Summit on AI for American industry in May 2018, bringing together industry stakeholders, academics and government leaders. Participants discussed the importance of removing barriers to AI innovation in the United States and promoting AI R&D collaboration among American allies. Participants also raised the need to promote awareness of AI so that the public can better understand how these technologies work and how they can benefit our daily lives. In the same month, it published a fact sheet called *Artificial Intelligence for the American People* (EOP, 2018[43]), which listed AI-related policies and measures by the current administration. These policies include increased public finance to AI R&D and regulatory reform to facilitate the development and use of drones and driverless cars. They also prioritise education in science, technology, engineering and mathematics (STEM) with a focus on computer science, as well as enhanced sharing of federal data for AI research and applications.

The president's FY 2019 and FY 2020 R&D budgets designated AI and machine learning as key priorities. Specific areas included basic AI research at the National Science Foundation and applied R&D at the Department of Transportation. Research priorities also include advanced health analytics at the National Institutes of Health and AI computing infrastructure at the Department of Energy. Overall, the federal government's investment in unclassified R&D for AI and related technologies has grown by over 40% since 2015.

In September 2018, the Select Committee on AI of the National Science and Technology Council began updating the National Artificial Intelligence Research and Development Strategic Plan. Since the plan's publication in 2016, the underlying technology, use cases and commercial implementation of AI have advanced rapidly. The Select Committee is seeking public input on how to improve the plan, including from those directly performing AI R&D and those affected by it.

The Administration has also prioritised training the future American workforce. President Trump signed an Executive Order establishing industry-recognised apprenticeships and creating a cabinet-level Task Force on Apprenticeship Expansion. In keeping with the policies outlined in the fact sheet, a Presidential Memorandum also prioritised high-quality STEM education, with a particular focus on computer science education. It committed USD 200 million in grant funds that were matched by a private industry commitment of USD 300 million.

The US Congress launched the bipartisan Artificial Intelligence Caucus in May 2017, which is co-chaired by congressmen John K. Delaney and Pete Olson (US, 2017[44]). The caucus brings together experts from academia, government and the private sector to discuss the implications of AI technologies. The Congress is considering legislation to establish both a federal AI advisory committee and federal safety standards for self-driving vehicles.

## Intergovernmental initiatives

### G7 and G20

At the April 2016 G7 ICT Ministerial Meeting of Takamatsu (Japan), the Japanese Minister of Internal Affairs and Communications presented and discussed a set of AI R&D Principles (G7, 2016[45]).

The G7 ICT and Industry Ministerial held in Turin in September 2017 under the Italian presidency issued a Ministerial Declaration in which G7 countries acknowledged the tremendous potential benefits of AI on society and economy, and agreed on a human-centred approach to AI (G7, 2017[46]).

Under Canada's 2018 G7 Presidency, G7 Innovation Ministers convened in Montreal in March 2018. They expressed a vision of human-centred AI and focused on the interconnected relationship between supporting economic growth from AI innovation. They also sought to increase trust in and adoption of AI, and promote inclusivity in AI development and deployment. G7 members agreed to act in related areas, including the following:

- Invest in basic and early-stage applied R&D to produce AI innovations, and support entrepreneurship in AI and labour force readiness for automation.

- Continue to encourage research, including solving societal challenges, advancing economic growth, and examining ethical considerations of AI, as well as broader issues such as those related to automated decision making.

- Support public awareness efforts to communicate actual and potential benefits, and broader implications, of AI.

- Continue to advance appropriate technical, ethical and technologically neutral approaches.

- Support the free flow of information through the sharing of best practices and use cases on the provision of open, interoperable and safe access to government data for AI.

- Disseminate this G7 statement globally to promote AI development and collaboration in the international arena (G7, 2018[47]).

In Charlevoix in June 2018, the G7 released a communique to promote human-centred AI and commercial adoption of AI. They also agreed to continue advancing appropriate technical, ethical and technologically neutral approaches.

G7 Innovation Ministers decided to convene a multi-stakeholder conference on AI hosted by Canada in December 2018. It planned to discuss how to harness the positive transformational potential of AI to promote inclusive and sustainable economic growth. France was also expected to propose AI-related initiatives as part of its G7 Presidency in 2019.

G20 attention to AI can also be noted, particularly with discussion proposed by Japan under its upcoming G20 presidency in 2019 (G20, 2018[1]). The G20 Digital Economy Ministerial Meeting of Salta in 2018 notably encouraged "countries to enable individuals and businesses to benefit from digitalisation and emerging technologies", such as 5G, IoT and AI. It encouraged Japan, in its role as president during 2019, to continue the G20 work accomplished in 2018, among other priorities including AI.

## OECD

### OECD Principles for Trust in and Adoption of AI

In May 2018, the OECD's Committee on Digital Economy Policy established an Expert Group on Artificial Intelligence in Society. It was created to scope principles for public policy and international co-operation that would foster trust in and adoption of AI. Ultimately, these principles became the basis for the OECD *Recommendation of the Council on Artificial Intelligence* (OECD, 2019[3]), to which forty countries adhered on 22 May 2019. In the same spirit, the 2018 Ministerial Council Meeting Chair urged "the OECD to pursue multi-stakeholder discussions on the possible development of principles that should underpin the development and ethical application of artificial intelligence in the service of people".

The group consisted of more than 50 experts from different sectors and disciplines. These included governments, business, the technical community, labour and civil society, as well as the European Commission and UNESCO. It held four meetings: two at the OECD in Paris, on 24-25 September and 12 November 2018; one at the Massachusetts Institute of Technology (MIT) in Cambridge on 16-17 January 2019; and a final meeting in Dubai, on 8-9 February 2019, on the margins of the World Government Summit. The group identified principles for the responsible stewardship of trustworthy AI relevant for all stakeholders. These principles included respect for human rights, fairness, transparency and explainability, robustness and safety, and accountability. The group also proposed specific recommendations for national policies to implement the principles. This work informed the development of the OECD *Recommendation of the Council on Artificial Intelligence* in the first half of 2019 (OECD, 2019[3]).

*OECD AI Policy Observatory*

The OECD planned to launch an AI Policy Observatory in 2019 to examine current and prospective developments in AI and their policy implications. The aim was to help implement the aforementioned AI principles by working with a wide spectrum of external stakeholders, including governments, industry, academia, technical experts and the general public. The Observatory was expected to be a multidisciplinary, evidence-based and multi-stakeholder centre for policy-relevant evidence collection, debate and guidance for governments. At the same time, it would provide external partners with a single window onto policy-relevant AI activities and findings from across the OECD.

## European Commission and other European institutions

In April 2018, the European Commission issued a Communication on Artificial Intelligence for Europe, outlining three main priorities. First, it boosts the European Union's technological and industrial capacity and AI uptake across the economy. Second, it prepares for socio-economic changes brought by AI. Third, it ensures an appropriate ethical and legal framework. The Commission presented a co-ordinated plan on the development of AI in Europe in December 2018. It aims primarily to maximise the impact of investments and collectively define the way forward. The plan was expected to run until 2027 and contains some 70 individual measures in the following areas:

- **Strategic actions and co-ordination**: encouraging member states to set up national AI strategies outlining investment levels and implementation measures.

- **Maximising investments through partnerships**: fostering investment in strategic AI research and innovation through AI public-private partnerships and a Leaders' Group, as well as a specific fund to support AI start-ups and innovative SMEs.

- **From the lab to the market**: strengthening research excellence centres and Digital Innovation Hubs, and establishing testing facilities and possibly regulatory sandboxes.

- **Skills and lifelong learning**: promoting talent, skills and lifelong learning.

- **Data**: calling for the creation of a Common European Data Space to facilitate access to data of public interest and industrial data platforms for AI, including health data.

- **Ethics by design and regulatory framework**: emphasising the need for ethical AI and for a regulatory framework fit for purpose (including safety and liability dimensions). The ethical framework will build on the AI Ethics Guidelines developed by the independent AI HLEG. The EC also commits to anchoring "ethics by design" through its procurement policy.

- **AI for the Public Sector**: outlining measures for AI for the public sector such as joint procurement and translations.

- **International co-operation**: underlining the importance of international outreach and of anchoring AI in development policy and announcing a world ministerial meeting in 2019.

As part of its AI Strategy, the Commission also established the AI HLEG in June 2018. The AI HLEG, which comprises representatives from academia, civil society and industry, was given two tasks. First, it was to draft AI Ethics Guidelines providing guidance to developers, deployers and users to ensure "trustworthy AI". Second, it was to prepare AI policy and investment recommendations ("Recommendations") for the European Commission and member states on mid- to long-term AI-related developments to advance Europe's global competitiveness. In parallel, the Commission set up a multi-stakeholder forum, the European AI Alliance, to encourage broad discussions on AI policy in Europe. Anyone can contribute through the platform to the work of the AI HLEG and inform EU policy making.

The AI HLEG published the first draft of its *Ethical Guidelines* for comments in December 2018. The draft guidelines constitute a framework to achieve trustworthy AI grounded in EU fundamental rights. To be trustworthy, AI should be lawful, ethical and socio-technically robust. The guidelines set out a set of ethical principles for AI. They also identify key requirements to ensure trustworthy AI, and methods to implement these requirements. Finally, they contain a non-exhaustive assessment list to help translate each requirement into practical questions that can help stakeholders put the principles into action. At the time of writing, the AI HLEG was revising the guidelines in view of comments, for official presentation to the EC on 9 April 2019. The EC was to explain the next steps for the guidelines and for a global ethical framework for AI. The second deliverable of the AI HLEG, the Recommendations, was due by the summer of 2019.

In 2017, the Council of Europe (CoE)'s Parliamentary Assembly published a Recommendation on technological convergence, AI and human rights. The Recommendation urged the Committee of Ministers to instruct CoE bodies to consider how emerging technologies such as AI challenge human rights. It also called for guidelines on issues such as transparency, accountability and profiling. In February 2019, the CoE's Committee of Ministers adopted a declaration on the manipulative capabilities of algorithmic processes. The declaration recognised "dangers for democratic societies" that arise from the capacity of "machine-learning tools to influence emotions and thoughts" and encouraging member states to address this threat. In February 2019, the CoE held a high-level conference called "Governing the Game Changer – Impacts of Artificial Intelligence Development on Human Rights, Democracy and the Rule of Law".

In addition, the CoE's European Commission for the Efficiency of Justice adopted the first European Ethical Charter on the use of AI in judicial systems in December 2018. It set out five principles to guide development of AI tools in European judiciaries. In 2019, the Committee on Legal Affairs and Human Rights decided to create a subcommittee on AI and human rights.

In May 2017, the European Economic and Social Committee (EESC) adopted an opinion on the societal impact of AI. The opinion called on EU stakeholders to ensure that AI development, deployment and use work for society and social well-being. The EESC said humans should keep control over when and how AI is used in daily lives, and identified 12 areas where AI raises societal concerns. These areas include ethics, safety, transparency, privacy, standards, labour, education, access, laws and regulations, governance, democracy, but also warfare and superintelligence. The opinion called for pan-European standards for AI ethics, adapted labour strategies and a European AI infrastructure with open-source learning environments (Muller, 2017[48]). An EESC temporary study group on AI has been set up to look at these issues.

### Nordic-Baltic region

In May 2018, ministers from Nordic and Baltic countries signed a joint declaration "AI in the Nordic-Baltic Region". The countries within the region comprise Denmark, Estonia, Finland, the Faroe Islands, Iceland, Latvia, Lithuania, Norway, Sweden and the Åland Islands. Together, they agreed to reinforce their co-operation on AI, while maintaining their position as Europe's leading region in the area of digital development (Nordic, 2018[49]).

The declaration identified seven focus areas to develop and promote the use of AI to better serve people. First, the countries want to improve opportunities for skills development so that more authorities, companies and organisations use AI. Second, they plan to enhance access to data for AI to be used for better services to citizens and businesses in the region. Third, they intend to develop ethical and transparent guidelines, standards, principles and values to guide when and how AI applications should be used. Fourth, they want infrastructure, hardware, software and data, all of which are central to the use of AI, to be based on standards that enable interoperability, privacy, security, trust, good usability and portability. Fifth, the countries will ensure that AI gets a prominent place in the European discussion and implementation of initiatives within the framework of the Digital Single Market. Sixth, they will avoid unnecessary regulation in the area, which is under rapid development. Seventh, they will use the Nordic Council of Ministers to facilitate collaboration in relevant policy areas.

### United Nations

In September 2017, the United Nations Interregional Crime and Justice Research Institute signed the Host Country Agreement to open a Centre on Artificial Intelligence and Robotics within the UN system in The Hague, The Netherlands.[2]

The International Telecommunication Union worked with more than 25 other UN agencies to host the "AI for Good" Global Summit. It also partnered with organisations such as the XPRIZE Foundation and the Association for Computing Machinery. Following a first summit in June 2017, the International Telecommunication Union held a second one in Geneva in May 2018.[3]

UNESCO has launched a global dialogue on the ethics of AI due to its complexity and impact on society and humanity. It held a public roundtable with experts in September 2018, as well as a global conference entitled "Principles for AI: Towards a Humanistic Approach? – AI with Human Values for Sustainable Development" in March 2019. Together, they aimed to raise awareness and promote reflection on the opportunities and challenges posed by AI and related technologies. In November 2019, UNESCO's 40th General Conference was to consider development of a recommendation on AI in 2020-21, if approved by UNESCO's Executive Board in April 2019.

### International Organization for Standardization

The International Organization for Standardization (ISO) and the International Electrotechnical Commission (IEC) created a joint technical committee ISO/IEC JTC 1 in 1987. They tasked the committee to develop information technology standards for business and consumer applications. In October 2017, subcommittee 42 (SC 42) was set up under JTC 1 to develop AI standards. SC 42 provides guidance to ISO and IEC committees developing AI applications. Its activities include providing a common framework and vocabulary, identifying computational approaches and architectures of AI systems, and evaluating threats and risks associated to them (Price, 2018[50]).

## Private stakeholder initiatives

Non-government stakeholders have formed numerous partnerships and initiatives to discuss AI issues. While many of these initiatives are multi-stakeholder in nature, this section primarily describes those based in the technical community, the private sector, labour or academia. This list is not exhaustive.

## Technical community and academia

The IEEE launched its Global Initiative on Ethics of Autonomous and Intelligent Systems in April 2016. This aims to advance public discussion on implementation of AI technologies and to define priority values and ethics. The IEEE published version 2 of its Ethically Aligned Design principles in December 2017, inviting comments from the public. It planned to publish the final version of the design guidelines in 2019 (Table 5.3) (IEEE, 2017[51]). Together with the MIT Media Lab, the IEEE launched a Council for Extended Intelligence in June 2018. It seeks to foster responsible creation of intelligent systems, reclaim control over personal data and create metrics of economic prosperity other than gross domestic product (Pretz, 2018[52]).

**Table 5.3. General Principles contained in the IEEE's Ethically Aligned Design (version 2)**

| Principles | Objectives |
|---|---|
| Human rights | Ensure autonomous and intelligent systems (AISs) do not infringe on internationally recognised human rights |
| Prioritising well-being | Prioritise metrics of well-being in the design and use of AISs because traditional metrics of prosperity do not take into account the full effect of AI systems technologies on human well-being |
| Accountability | Ensure that designers and operators of AISs are responsible and accountable |
| Transparency | Ensure AIS operate in a transparent manner |
| AIS technology misuse and awareness of it | Minimise the risks of misuse of AIS technology |

*Source*: IEEE (2017[51]), *Ethically Aligned Design (Version 2)*, http://standards.ieee.org/develop/indconn/ec/ead_v2.pdf.

The Asilomar AI Principles are 23 principles for the safe and socially beneficial development of AI in the near and longer term that resulted from the Future Life Institute's conference of January 2017. The Asilomar conference extracted core principles from discussions, reflections and documents produced by the IEEE, academia and non-profit organisations.

The issues are grouped into three areas. Research issues call for research funding for beneficial AI that include difficult questions in computer science; economics, law and social studies; a constructive "science-policy link"; and a technical research culture of co-operation, trust and transparency. Ethics and values call for AI systems' design and operation to be safe and secure, transparent and accountable, protective of individuals' liberty, privacy, human dignity, rights and cultural diversity, broad empowerment and shared benefits. Longer-term issues notably avoid strong assumptions on the upper limits of future AI capabilities and plan carefully for the possible development of artificial general intelligence (AGI) (FLI, 2017[53]). Table 5.4 provides the list of Asilomar AI Principles.

The non-profit AI research company OpenAI was founded in late 2015. It employs 60 full-time researchers with the mission to "build safe AGI, and ensure AGI's benefits are as widely and evenly distributed as possible".[4]

The AI Initiative was created by the Future Society in 2015 to help shape the global AI policy framework. It hosts an online platform for multidisciplinary civic discussion and debate. This platform aims to help understand the dynamics, benefits and risks of AI technology so as to inform policy recommendations.[5]

There are also numerous academic initiatives in all OECD countries and many partner economies. The MIT Internet Policy Research Initiative, for example, is helping bridge the gap between technical and policy communities. The Berkman Klein Center at Harvard University launched the Ethics and Governance of Artificial Intelligence Initiative in 2017.

For its part, the MIT Media Lab is focusing on algorithms and justice, autonomous vehicles and the transparency and explainability of AI.

**Table 5.4. Asilomar AI Principles (excerpt titles of the principles)**

| | Research issues | Ethics and values | Longer-term issues |
|---|---|---|---|
| Titles of principles | - Research goal<br>- Research funding<br>- Research funding<br>- Science-policy link<br>- Research culture<br>- Race avoidance | - Safety<br>- Failure transparency<br>- Judicial transparency<br>- Responsibility<br>- Value alignment<br>- Human values<br>- Personal privacy<br>- Liberty and privacy<br>- Shared benefit<br>- Shared prosperity<br>- Human control | - Capability caution<br>- Importance<br>- Risks<br>- Recursive self-improvement<br>- Common good |

*Source*: FLI (2017[53]), *Asilomar AI Principles*, https://futureoflife.org/ai-principles/.

## Private-sector initiatives

In September 2016, Amazon, DeepMindGoogle, Facebook, IBM and Microsoft launched the Partnership on Artificial Intelligence to Benefit People and Society (PAI). It aims to study and formulate best practices on AI technologies, advance the public's understanding of AI and serve as an open platform for discussion and engagement about AI and its influences on people and society. Since its creation, PAI has become a multidisciplinary stakeholder community with more than 80 members. They range from for-profit technology companies to representatives of civil society, academic and research institutions, and start-ups.

The Information Technology Industry Council (ITI) is a business association of technology companies based in Washington, DC with more than 60 members. ITI published *AI Policy Principles* in October 2017 (Table 5.5). The principles identified industry's responsibility in certain areas, and called for government support of AI research and for public-private partnerships (ITI, 2017[54]). Companies are taking action individually too.

**Table 5.5. ITI AI Policy Principles**

| Responsibility: Promoting responsible development and use | Opportunity for governments: Investing and enabling the AI ecosystem | Opportunity for public-private partnerships: Promoting lifespan education and diversity |
|---|---|---|
| - Responsible design and deployment<br>- Safety and controllability<br>- Robust and representative data<br>- Interpretability<br>- Liability of AI systems due to autonomy | - Investment in AI research and development<br>- Flexible regulatory approach<br>- Promoting innovation and the security of the Internet<br>- Cybersecurity and privacy<br>- Global standards and best practices | - Democratising access and creating equality of opportunity<br>- Science, technology, engineering and mathematics education<br>- Workforce<br>- Public-private partnership |

*Source*: ITI (2017[54]), *AI Policy Principles*, https://www.itic.org/resources/AI-Policy-Principles-FullReport2.pdf.

## Civil society

The Public Voice coalition, established by the Electronic Privacy Information Center, published *Universal Guidelines on Artificial Intelligence* (UGAI) in October 2018 (The Public Voice,

2018[55]). The guidelines call attention to the growing challenges of intelligent computational systems and propose concrete recommendations to improve and inform their design. At its core, the UGAI promotes transparency and accountability of AI systems and seeks to ensure that people retain control over the systems they create.[6] The 12 UGAI Principles address various rights and obligations. These comprise the right to transparency and human determination, and obligations to identification; fairness; assessment and accountability; accuracy, reliability and validity; data quality; public safety; cybersecurity; and termination. They also include prohibitions on secret profiling and unitary scoring.

## *Labour organisations*

UNI Global Union represents more than 20 million workers from over 150 countries in skills and services sectors. A future that empowers workers and provides decent work is a key UNI Global Union priority. It has identified ten key principles for Ethical AI. These aim to ensure that collective agreements, global framework agreements and multinational alliances involving unions, shop stewards and global alliances respect workers' rights (Table 5.6) (Colclough, 2018[56]).

**Table 5.6. Top 10 Principles for Ethical Artificial Intelligence (UNI Global Union)**

| | |
|---|---|
| 1. AI systems must be transparent: | Workers should have the right to demand transparency in the decisions and outcomes of AI systems, as well as their underlying algorithms. They must also be consulted on AI systems implementation, development and deployment. |
| 2. AI systems must be equipped with an ethical black box: | The ethical black box should not only contain relevant data to ensure system transparency and accountability, but also clear data and information on the ethical considerations built into the system. |
| 3. AI must serve people and planet: | Codes of ethics for the development, application and use of AI are needed so that throughout their entire operational process, AI systems remain compatible and increase the principles of human dignity, integrity, freedom, privacy, and cultural and gender diversity, as well as fundamental human rights. |
| 4. Adopt a human-in-command approach: | The development of AI must be responsible, safe and useful, where machines maintain the legal status of tools, and legal persons retain control over, and responsibility for, these machines at all times. |
| 5. Ensure a genderless, unbiased AI: | In the design and maintenance of AI and artificial systems, it is vital that the system is controlled for negative or harmful human-bias, and that any bias be it gender, race, sexual orientation or age is identified and is not propagated by the system. |
| 6. Share the benefits of AI systems: | The economic prosperity created by AI should be distributed broadly and equally, to benefit all of humanity. Global as well as national policies aimed at bridging the economic, technological and social digital divide are therefore necessary. |
| 7. Secure a just transition and ensure support for fundamental freedoms and rights: | As AI systems develop and augmented realities are formed, workers and work tasks will be displaced. It is vital that policies are put in place that ensure a just transition to the digital reality, including specific governmental measures to help displaced workers find new employment. |
| 8. Establish global governance mechanism: | Establish multi-stakeholder Decent Work and Ethical AI governance bodies on global and regional levels. The bodies should include AI designers, manufacturers, owners, developers, researchers, employers, lawyers, civil society organisations and trade unions. |
| 9. Ban the attribution of responsibility to robots: | Robots should be designed and operated as far as is practicable to comply with existing laws, and fundamental rights and freedoms, including privacy. |
| 10. Ban AI arms race: | Lethal autonomous weapons, including cyber warfare, should be banned. UNI Global Union calls for a global convention on ethical AI that will help address, and work to prevent, the unintended negative consequences of AI while accentuating its benefits to workers and society. We underline that humans and corporations are the responsible agents. |

*Source*: Colclough (2018[56]), "Ethical Artificial Intelligence – 10 Essential Ingredients", https://www.oecd-forum.org/channels/722-digitalisation/posts/29527-10-principles-for-ethical-artificial-intelligence.

## References

Benhamou, S. and L. Janin (2018), *Intelligence artificielle et travail*, France Stratégie, [21]
http://www.strategie.gouv.fr/publications/intelligence-artificielle-travail.

Brazil (2018), *Brazilian digital transformation strategy "E-digital"*, Ministério da Ciência, [5]
Tecnologia, Inovações e Comunicações,
http://www.mctic.gov.br/mctic/export/sites/institucional/sessaoPublica/arquivos/digitalstrateg
y.pdf.

China (2018), *AI Standardisation White Paper*, Government of China, translated into English by [12]
Jeffrey Ding, Researcher in the Future of Humanity's Governance of AI Program,
https://baijia.baidu.com/s?id=1589996219403096393.

China (2017), *Guideline on Next Generation AI Development Plan*, Government of China, State [10]
Council, http://www.gov.cn/zhengce/content/2017-07/20/content_5211996.htm.

China (2016), *Three-Year Action Plan for Promoting Development of a New Generation* [9]
*Artificial Intelligence Industry (2018-2020)*, Chinese Ministry of Industry and Information
Technology, http://www.miit.gov.cn/n1146290/n1146392/c4808445/content.html.

CIFAR (2017), *Pan-Canadian Artificial Intelligence Strategy*, CIFAR, [6]
https://www.cifar.ca/ai/pan-canadian-artificial-intelligence-strategy.

Colclough, C. (2018), "Ethical Artificial Intelligence – 10 Essential Ingredients", *A.Ideas Series*, [56]
No. 24, The Forum Network, OECD, Paris, https://www.oecd-forum.org/channels/722-
digitalisation/posts/29527-10-principles-for-ethical-artificial-intelligence.

Denmark (2018), *Strategy for Denmark's Digital Growth*, Government of Denmark, [14]
https://em.dk/english/publications/2018/strategy-for-denmarks-digital-growth.

EOP (2018), *Artificial Intelligence for the American People*, Executive Office of the President, [43]
Government of the United States, https://www.whitehouse.gov/briefings-statements/artificial-
intelligence-american-people/.

Finland (2017), *Finland's Age of Artificial Intelligence - Turning Finland into a Leader in the* [17]
*Application of AI*, webpage, Finnish Ministry of Economic Affairs and Employment,
https://tem.fi/en/artificial-intelligence-programme.

FLI (2017), *Asilomar AI Principles*, Future of Life Institute (FLI), https://futureoflife.org/ai- [53]
principles/.

Fonds de recherche du Québec (2018), "Québec lays the groundwork for a world observatory on [7]
the social impacts of artificial intelligence and digital technologies", News Release, 29
March, https://www.newswire.ca/news-releases/quebec-lays-the-groundwork-for-a-world-
observatory-on-the-social-impacts-of-artificial-intelligence-and-digital-technologies-
678316673.html.

G20 (2018), *Ministerial Declaration – G20 Digital Economy*, G20 Digital Economy Ministerial [1]
Meeting, 24 August, Salta, Argentina,
https://www.g20.org/sites/default/files/documentos_producidos/digital_economy_-
_ministerial_declaration_0.pdf.

G7 (2018), *Chairs' Summary: G7 Ministerial Meeting on Preparing for Jobs of the Future*, [47]
https://g7.gc.ca/en/g7-presidency/themes/preparing-jobs-future/.

G7 (2017), *Artificial Intelligence (Annex 2)*,
http://www.g7italy.it/sites/default/files/documents/ANNEX2-Artificial_Intelligence_0.pdf. [46]

G7 (2016), *Proposal of Discussion toward Formulation of AI R&D Guideline*, Japanese Ministry [45]
of Internal Affairs and Communications,
http://www.soumu.go.jp/joho_kokusai/g7ict/english/index.html.

Germany (2018), *Artificial Intelligence Strategy*, Federal Government of Germany, [22]
https://www.ki-strategie-deutschland.de/home.html.

Germany (2017), *Automated and Connected Driving*, BMVI Ethics Commission, [23]
https://www.bmvi.de/SharedDocs/EN/publications/report-ethics-
commission.pdf?__blob=publicationFile.

Hall, W. and J. Pesenti (2017), *Growing the Artificial Intelligence Industry in the UK*, Wendy [39]
Hall and Jérôme Pesenti,
https://assets.publishing.service.gov.uk/government/uploads/system/uploads/attachment_data/
file/652097/Growing_the_artificial_intelligence_industry_in_the_UK.pdf.

IEEE (2017), *Ethically Aligned Design (Version 2)*, IEEE Global Initiative on Ethics of [51]
Autonomous and Intelligent Systems,
http://standards.ieee.org/develop/indconn/ec/ead_v2.pdf.

India (2018), "National Strategy for Artificial Intelligence #AI for All", Discussion Paper, NITI [24]
Aayog, http://niti.gov.in/writereaddata/files/document_publication/NationalStrategy-for-AI-
Discussion-Paper.pdf.

Italy (2018), *Artificial Intelligence at the Service of the Citizen*, Agency for Digital Italy, [25]
https://libro-bianco-ia.readthedocs.io/en/latest/.

ITI (2017), *AI Policy Principles*, Information Technology Industry Council, [54]
https://www.itic.org/resources/AI-Policy-Principles-FullReport2.pdf.

Japan (2019), *Social Principles for Human-Centric AI*, Japan Cabinet Office, April, [28]
https://www8.cao.go.jp/cstp/stmain/aisocialprinciples.pdf.

Japan (2018), *Draft AI Utilization Principles*, Ministry of Internal Affairs and Communications, [30]
Japan, http://www.soumu.go.jp/main_content/000581310.pdf.

Japan (2018), *Integrated Innovation Strategy*, Japan Cabinet Office, June, [27]
https://www8.cao.go.jp/cstp/english/doc/integrated_main.pdf.

Japan (2017), *Artificial Intelligence Technology Strategy*, Strategic Council for AI Technology, [26]
http://www.nedo.go.jp/content/100865202.pdf.

Japan (2017), *Draft AI R&D Guidelines for International Discussions*, Ministry of Internal [29]
Affairs and Communications, Japan, http://www.soumu.go.jp/main_content/000507517.pdf.

Jing, M. and S. Dai (2018), *Here's what China is doing to boost its artificial intelligence* [8]
*capabilities*, 10 May, https://www.scmp.com/tech/science-research/article/2145568/can-
trumps-ai-summit-match-chinas-ambitious-strategic-plan.

Kaevats, M. (2017), *Estonia's Ideas on Legalising AI*, presentation at the "AI: Intelligent [15]
Machines, Smart Policies" conference, Paris, 26-27 October,
https://prezi.com/yabrlekhmcj4/oecd-6-7min-paris/.

Kaevats, M. (25 September 2017), *Estonia considers a 'kratt law' to legalise artifical intelligence (AI)*, E-residency blog, https://medium.com/e-residency-blog/estonia-starts-public-discussion-legalising-ai-166cb8e34596. [16]

Kania, E. (2018), "China's AI agenda advances", *The Diplomat,* 14 February, https://thediplomat.com/2018/02/chinas-ai-agenda-advances/. [11]

Korea (2016), *Mid- to Long-term Master Plan in Preparation for the Intelligent Information Society*, Government of Korea, Interdepartmental Exercise, http://english.msip.go.kr/cms/english/pl/policies2/__icsFiles/afieldfile/2017/07/20/Master%20Plan%20for%20the%20intelligent%20information%20society.pdf. [32]

Korea (2016), "MSIP announces development strategy for the intelligence information industry", *Science, Technology & ICT Newsletter, Ministry of Science and ICT, Government of Korea*, No. 16, https://english.msit.go.kr/english/msipContents/contentsView.do?cateId=msse44&artId=1296203. [31]

Mexico (2018), *Towards an AI strategy in Mexico: Harnessing the AI Revolution*, British Embassy Mexico City, Oxford Insights, C minds, http://go.wizeline.com/rs/571-SRN-279/images/Towards-an-AI-strategy-in-Mexico.pdf. [34]

Muller, C. (2017), *Opinion on the Societal Impact of AI*, European Economic and Social Committee, Brussels, https://www.eesc.europa.eu/en/our-work/opinions-information-reports/opinions/artificial-intelligence. [48]

Nordic (2018), *AI in the Nordic-Baltic Region*, Nordic Council of Ministers, https://www.regeringen.se/49a602/globalassets/regeringen/dokument/naringsdepartementet/20180514_nmr_deklaration-slutlig-webb.pdf. [49]

OECD (2019), *Recommendation of the Council on Artificial Intelligence*, OECD, Paris. [3]

OECD (2019), *Scoping Principles to Foster Trust in and Adoption of AI – Proposal by the Expert Group on Artificial Intelligence at the OECD (AIGO)*, OECD, Paris, http://oe.cd/ai. [2]

OGCR (2018), *Analysis of the Development Potential of Artificial Intelligence in the Czcech Republic*, Office of the Government of the Czcech Republic, https://www.vlada.cz/assets/evropske-zalezitosti/aktualne/AI-Summary-Report.pdf. [13]

Peng, T. (2018), "South Korea aims high on AI, pumps $2 billion into R&D", *Medium,* 16 May, https://medium.com/syncedreview/south-korea-aims-high-on-ai-pumps-2-billion-into-r-d-de8e5c0c8ac5. [33]

Porter, M. (1990), "The competitive advantage of nations", *Harvard Business Review,* March-April, https://hbr.org/1990/03/the-competitive-advantage-of-nations. [4]

Pretz, K. (2018), "IEEE Standards Association and MIT Media Lab form council on extended intelligence", *IEEE Spectrum*, http://theinstitute.ieee.org/resources/ieee-news/ieee-standards-association-and-mit-media-lab-form-council-on-extended-intelligence. [52]

Price, A. (2018), "First international standards committee for entire AI ecosystem", *IE e-tech,* Issue 03, https://iecetech.org/Technical-Committees/2018-03/First-International-Standards-committee-for-entire-AI-ecosystem. [50]

Russia (2017), *Digital Economy of the Russian Federation*, Government of the Russian Federation, http://pravo.gov.ru. [35]

Singapore (2018), *Digital Economy Framework for Action*, Infocomm Media Development Authority, https://www.imda.gov.sg/-/media/imda/files/sg-digital/sgd-framework-for-action.pdf?la=en.     [36]

Sivonen, P. (2017), *Ambitious Development Program Enabling Rapid Growth of AI and Platform Economy in Finland*, presentation at the "AI Intelligent Machines, Smart Policies" conference, Paris, 26-27 October, http://www.oecd.org/going-digital/ai-intelligent-machines-smart-policies/conference-agenda/ai-intelligent-machines-smart-policies-sivonen.pdf.     [18]

The Public Voice (2018), *Universal Guidelines for Artificial Intelligence*, The Public Voice Coalition, October, https://thepublicvoice.org/ai-universal-guidelines/memo/.     [55]

Thompson, N. (2018), "Emmanuel Macron talks to WIRED about France's AI strategy", *WIRED*, 31 March, https://www.wired.com/story/emmanuel-macron-talks-to-wired-about-frances-ai-strategy.     [20]

UK (2018), *AI Sector Deal*, Department for Business, Energy & Industrial Strategy and Department for Digital, Culture, Media & Sport, Government of the United Kingdom, https://www.gov.uk/government/publications/artificial-intelligence-sector-deal.     [41]

UK (2018), *Centre for Data Ethics and Innovation Consultation*, Department for Digital, Culture, Media & Sport, Government of the United Kingdom, https://www.gov.uk/government/consultations/consultation-on-the-centre-for-data-ethics-and-innovation/centre-for-data-ethics-and-innovation-consultation.     [42]

UK (2017), *Industrial Strategy: Building a Britain Fit for the Future*, Government of the United Kingdom, https://www.gov.uk/government/publications/industrial-strategy-building-a-britain-fit-for-the-future.     [40]

UK (2017), *UK Digital Strategy*, Government of the United Kingdom, https://www.gov.uk/government/publications/uk-digital-strategy/uk-digital-strategy.     [38]

US (2017), "Delaney launches bipartisan artificial intelligence (AI) caucus for 115th Congress", Congressional Artificial Intelligence Caucus, News Release, 24 May, https://artificialintelligencecaucus-olson.house.gov/media-center/press-releases/delaney-launches-ai-caucus.     [44]

Villani, C. (2018), *For a Meaningful Artificial Intelligence - Towards a French and European Strategy*, AI for Humanity, https://www.aiforhumanity.fr/ (accessed on 15 December 2018).     [19]

Vinnova (2018), *Artificial Intelligence in Swedish Business and Society*, Vinnova, 28 October, https://www.vinnova.se/contentassets/29cd313d690e4be3a8d861ad05a4ee48/vr_18_09.pdf.     [37]

## Notes

[1] The report was commissioned by the British Embassy in Mexico, funded by the United Kingdom's Prosperity Fund and developed by Oxford Insights and C Minds, with the collaboration of the Mexican Government and input from experts across Mexico.

[2] See www.unicri.it/news/article/2017-09-07_Establishment_of_the_UNICRI.

[3] See https://www.itu.int/en/ITU-T/AI/.

[4] See https://openai.com/about/#mission.

[5] See http://ai-initiative.org/ai-consultation/.

[6] For more information, see https://thepublicvoice.org/ai-universal-guidelines/memo/.

# ORGANISATION FOR ECONOMIC CO-OPERATION AND DEVELOPMENT

The OECD is a unique forum where governments work together to address the economic, social and environmental challenges of globalisation. The OECD is also at the forefront of efforts to understand and to help governments respond to new developments and concerns, such as corporate governance, the information economy and the challenges of an ageing population. The Organisation provides a setting where governments can compare policy experiences, seek answers to common problems, identify good practice and work to co-ordinate domestic and international policies.

The OECD member countries are: Australia, Austria, Belgium, Canada, Chile, the Czech Republic, Denmark, Estonia, Finland, France, Germany, Greece, Hungary, Iceland, Ireland, Israel, Italy, Japan, Korea, Latvia, Lithuania, Luxembourg, Mexico, the Netherlands, New Zealand, Norway, Poland, Portugal, the Slovak Republic, Slovenia, Spain, Sweden, Switzerland, Turkey, the United Kingdom and the United States. The European Union takes part in the work of the OECD.

OECD Publishing disseminates widely the results of the Organisation's statistics gathering and research on economic, social and environmental issues, as well as the conventions, guidelines and standards agreed by its members.

OECD PUBLISHING, 2, rue André-Pascal, 75775 PARIS CEDEX 16
ISBN 978-92-64-58254-5 – 2019